# TEST YOUR
# Higher CHEMISTRY
# CALCULATIONS

## John Anderson

SCOTTISH EXAMINATION MATERIALS

HODDER GIBSON

AN HACHETTE UK COMPANY

The Publishers would like to thank the following for permission to reproduce copyright material:

**Photo credits** p.9 © Imagestate Media (John Foxx)/Mouse SS136.

Every effort has been made to trace all copyright holders, but if any have been inadvertently overlooked the Publishers will be pleased to make the necessary arrangements at the first opportunity.

Although every effort has been made to ensure that website addresses are correct at time of going to press, Hodder Gibson cannot be held responsible for the content of any website mentioned in this book. It is sometimes possible to find a relocated web page by typing in the address of the home page for a website in the URL window of your browser.

Hachette UK's policy is to use papers that are natural, renewable and recyclable products and made from wood grown in sustainable forests. The logging and manufacturing processes are expected to conform to the environmental regulations of the country of origin.

Orders: please contact Bookpoint Ltd, 130 Park Drive, Milton Park, Abingdon, Oxon OX14 4SE. Telephone: (44) 01235 827720. Fax: (44) 01235 400454. Lines are open 9.00–5.00, Monday to Saturday, with a 24-hour message answering service. Visit our website at www.hoddereducation.co.uk. Hodder Gibson can be contacted direct on: Tel: 0141 333 4650; Fax: 0141 404 8188; email: hoddergibson@hodder.co.uk

© John Anderson 2017

First published in 2017 by
Hodder Gibson, an imprint of Hodder Education,
An Hachette UK Company,
211 St Vincent Street
Glasgow G2 5QY

Impression number   5  4  3  2  1
Year                 2021  2020  2019  2018  2017

Cover photo © CROWN COPYRIGHT/HEALTH & SAFETY LABORATORY/SCIENCE PHOTO LIBRARY
Illustrations by Aptara, Inc.
Typeset in Minion Pro 9/11pts by Aptara, Inc.
Printed in Italy

A catalogue record for this title is available from the British Library

ISBN: 978 1 4718 7385 0

# Contents

# Introduction

Calculations are an important part of the Higher Chemistry course. Being able to handle calculations with confidence is an essential skill, not only to achieve a good result in the Higher exam but to help you understand chemical processes. This book takes you through worked examples and then allows you to practise with lots of relevant questions. You can check your progress as you go using the answers provided at the end of the book.

This book is based on the original title *Test Your Higher Chemistry Calculations* which was written to support Higher Chemistry students prior to the introduction of the Curriculum for Excellence Higher. This title was highly successful as it presented calculations in an easy-to-understand style, provided relevant worked examples and gave students plenty of opportunity to refine their skills as each chapter contained several problems to solve, with answers provided.

This new book has retained the best features of the original title but the content has been updated to reflect the current Higher Chemistry curriculum. In addition, each chapter has been fully revised to bring examples up to date. Key features of this new book include:

- worked examples
- questions at the end of each chapter presented in order of increasing challenge
- new chapters on potential energy graphs, atom economy and bond enthalpy
- a chapter on numeracy to help students refine their numeracy skills
- answers to all problems.

This book complements the other study guides for Higher Chemistry published by Hodder Gibson. Further help, guidance and necessary explanations for the whole course can be gathered from these excellent textbooks:

*Higher Chemistry with Answers* by John Anderson, Eric Allan and John Harris

*Revision Notes and Questions for Higher Chemistry* by John Anderson

*How to Pass Higher Chemistry* by John Anderson

# Rates of reaction

Being able to control a chemical reaction is an essential skill for a chemist. Reactions which are too slow are usually unprofitable. Reactions which are too fast can be dangerous! Thus, chemists have discovered methods for controlling the rate of chemical reactions by, for example, changing the concentration of reactants or by adding a catalyst. In order to assess how effective these methods are at changing the rate, chemists need to know how to measure the rate of a chemical reaction.

## Measuring the rate of reaction

In previous work, you will have learned that chemists can measure the change in a reaction. For example, for a reaction producing a gas, it is common to measure the volume of gas produced or record the change in mass as the gas is 'lost' to the surroundings. Alternatively, since all chemical reactions involve reactants being changed into products, chemists can measure the change in the concentration of reactants or products.

Overall, the reaction rate can be calculated using the equation:

$$\text{rate} = \frac{\text{change in volume or mass or concentration}}{\text{time}}$$

## The relationship between rate and time

In Higher Chemistry, we are interested in the relationship between rate and time.

- A reaction which is over in a short time will have a high rate.

- A reaction which takes a long time to reach completion will have a low rate.

Mathematically, we say that the rate is inversely proportional to the time. This simply means that as the rate increases, the time will decrease. This can be expressed using the following equation:

$$\text{rate} = \frac{1}{\text{time}}$$

This is useful for measuring reactions where there is a visible sign that the reaction has ended. For example, in some reactions there is an obvious colour change when the reaction is complete. The rate can be measured by starting a timer as soon as the reactants are added and stopping the timer as soon as the colour change is observed.

### Using rate and time

The table below illustrates the numerical relationship between rate and time.

| Experiment | Time taken | Rate |
|---|---|---|
| 1 | 20 s | $0.05 \, s^{-1}$ |
| 2 | 100 s | $0.01 \, s^{-1}$ |
| 3 | 2 min | $0.5 \, min^{-1}$ |
| 4 | 50 min | $0.02 \, min^{-1}$ |

Points to note:

- The rate is calculated using the equation $\text{rate} = \frac{1}{\text{time}}$.

- The time is calculated using the equation $\text{time} = \frac{1}{\text{rate}}$.

- The units for rate are taken from the units for time. To show the inverse relationship, a superscript '−1' is added: $s^{-1}$, $min^{-1}$, etc.

- The standard unit of time is seconds so the standard unit for rate is $s^{-1}$. If you are given time in mixed units (e.g. 5 min 25 s) you should convert the time into seconds and calculate the rate using the unit $s^{-1}$. This is illustrated in the following worked examples.

## Worked example 1.1

Copy and complete the following table by calculating the rate of reaction for each experiment.

(Note that the reaction rate units are $s^{-1}$ so you will need to convert the time into seconds.)

| Experiment | Time taken | Rate of reaction/$s^{-1}$ |
|---|---|---|
| 1 | 10.2 s | a) |
| 2 | 303 s | b) |
| 3 | 8 min 35 s | c) |
| 4 | 12 min 24 s | d) |
| 5 | 1 hour 28 min 2 s | e) |

### Solution

a) Rate = $\frac{1}{10.2}$ = **0.098 $s^{-1}$**

b) Rate = $\frac{1}{303}$ = **0.0033 $s^{-1}$**

c) Time in seconds = (8 × 60) + 35 = 515 s; rate = $\frac{1}{515}$ = **0.0019 $s^{-1}$**

d) Time in seconds = (12 × 60) + 24 = 744 s; rate = $\frac{1}{744}$ = **0.0013 $s^{-1}$**

e) Time in seconds = (1 × 60 × 60) + (28 × 60) + 2 = 5282 s; rate = $\frac{1}{5282}$ = **1.89 × 10$^{-4}$ $s^{-1}$**

## Worked example 1.2

Copy and complete the following table by calculating the time taken for each reaction.

| Experiment | Rate of reaction/$s^{-1}$ | Time/s |
|---|---|---|
| 1 | 0.0039 | a) |
| 2 | 0.12 | b) |
| 3 | 2.3 × 10$^{-4}$ | c) |
| 4 | 5.54 × 10$^{-5}$ | d) |
| 5 | 1.39 × 10$^{-6}$ | e) |

### Solution

a) Time = $\frac{1}{0.0039}$ = **256.41 s**

b) Time = $\frac{1}{0.12}$ = **8.33 s**

c) Time = $\frac{1}{2.3 \times 10^{-4}}$ = **4348 s**

d) Time = $\frac{1}{5.54 \times 10^{-5}}$ = **18 050 s**

e) Time = $\frac{1}{1.39 \times 10^{-6}}$ = **719 424 s**

# Calculations from reaction rate graphs

In the Higher Chemistry exam, you are likely to encounter reaction rate calculations where the data are shown graphically. Tackling these questions is straightforward.

- Read the graph and extract the data.
- Apply the relevant equation: $\text{rate} = \dfrac{1}{\text{time}}$ or $\text{time} = \dfrac{1}{\text{rate}}$.
- Remember to show the correct units.

## Worked example 1.3

**Look at Figure 1.1.**

**a)** State the time taken when the concentration was
**(i)** 0.03 mol l$^{-1}$ **(ii)** 0.06 mol l$^{-1}$.

**b)** Calculate the concentration when the time taken was
**(i)** 500 s **(ii)** 86.2 s.

## Solution

**a) (i)** From the graph, at a concentration of 0.03 mol l$^{-1}$, the rate of reaction is 0.004 s$^{-1}$.
$$\text{time} = \frac{1}{\text{rate}} = \frac{1}{0.004} = \textbf{250 s}$$

**(ii)** From the graph, at a concentration of 0.06 mol l$^{-1}$, the rate of reaction is 0.008 s$^{-1}$.
$$\text{time} = \frac{1}{\text{rate}} = \frac{1}{0.008} = \textbf{125 s}$$

**b) (i)** The rate of reaction can be calculated using
$$\text{rate} = \frac{1}{\text{time}} = \frac{1}{500} = 0.002 \text{ s}^{-1}.$$
From the graph, this value for the rate of reaction occurs at a concentration of **0.015 mol l$^{-1}$**.

**(ii)** The rate of reaction can be calculated using
$$\text{rate} = \frac{1}{\text{time}} = \frac{1}{86.2} = 0.0116 \text{ s}^{-1}.$$
From the graph, this value for the rate of reaction occurs at a concentration of **0.086 mol l$^{-1}$**.

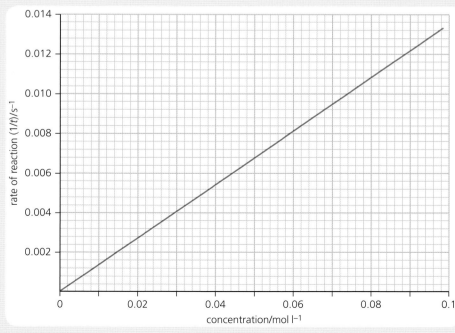

**Figure 1.1**

## Worked example 1.4

The graph on the right was obtained from carrying out an experiment at different temperatures and recording the time taken for a colour change to appear.

a) State the time taken when the reaction was carried out at **(i)** 20 °C **(ii)** 29 °C.

b) State the temperature of the experiment which took **(i)** 10 s **(ii)** 14.7 s.

### Solution

a) **(i)** From the graph, at a temperature of 20 °C, the rate of reaction is 0.04 s⁻¹.

$$\text{time} = \frac{1}{\text{rate}} = \frac{1}{0.04} = \mathbf{25\,s}$$

**(ii)** From the graph, at a temperature of 29 °C, the rate of reaction is 0.08 s⁻¹.

$$\text{time} = \frac{1}{\text{rate}} = \frac{1}{0.08} = \mathbf{12.5\,s}$$

b) **(i)** The rate of reaction can be calculated using $\text{rate} = \frac{1}{\text{time}} = \frac{1}{10} = 0.10\,s^{-1}$.
From the graph, this value for the rate of reaction occurs at **31 °C**.

**(ii)** The rate of reaction can be calculated using $\text{rate} = \frac{1}{\text{time}} = \frac{1}{14.7} = 0.068\,s^{-1}$.
From the graph, this value for the rate of reaction occurs at **27 °C**.

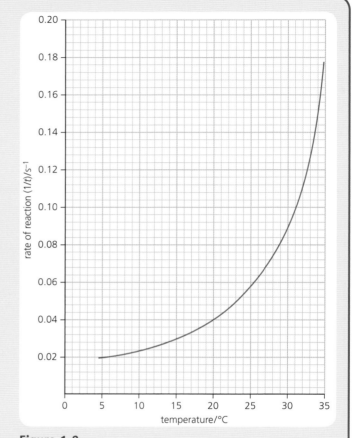

**Figure 1.2**

## Questions

1   Find out the values of a)–e) by calculating the rate, in $s^{-1}$, for each experiment.

| Experiment | Time | Rate/$s^{-1}$ |
|---|---|---|
| 1 | 23 s | a) |
| 2 | 380 s | b) |
| 3 | 585 s | c) |
| 4 | 2 min | d) |
| 5 | 4 min 20 s | e) |

2   Find out the values of a)–e) by calculating the time, in seconds, for each experiment.

| Experiment | Rate/$s^{-1}$ | Time/s |
|---|---|---|
| 1 | 0.012 | a) |
| 2 | 0.049 | b) |
| 3 | $1.1 \times 10^{-3}$ | c) |
| 4 | $5.65 \times 10^{-4}$ | d) |
| 5 | $6.75 \times 10^{-4}$ | e) |

3   Figure 1.3 shows how the rate of a reaction varies with concentration.

   a) State the concentration when the time taken was 20 s.

   b) State the time taken if the experiment was carried out using a concentration of 0.15 mol l$^{-1}$.

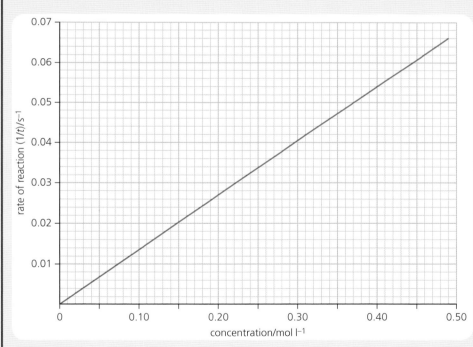

**Figure 1.3**

4 Figure 1.4 shows how the rate of a reaction varies with temperature.

a) State the time taken for the reaction at a temperature of 43 °C.

b) State the temperature of the experiment for the reaction which took 50 s.

c) Obtain from the graph the values for the rate of reaction at 10 °C, 20 °C, 30 °C and 40 °C. What relationship between the temperatures and the rates can be observed?

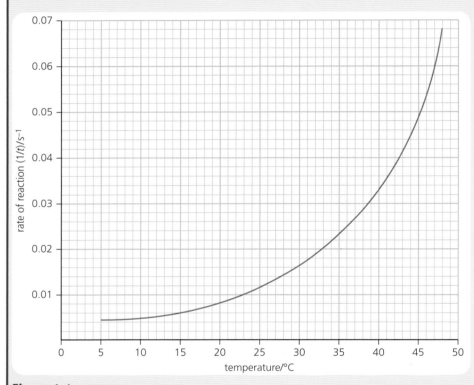

**Figure 1.4**

5   Figure 1.5 shows how the rate of a reaction varies with concentration. The time taken for a colour change to appear was used to measure the rate of the reaction.

a) What concentration of reactant will cause the colour change in this reaction to take place after 50 s?

b) After what time will the colour change take place if the reagent concentration is 0.40 mol l⁻¹?

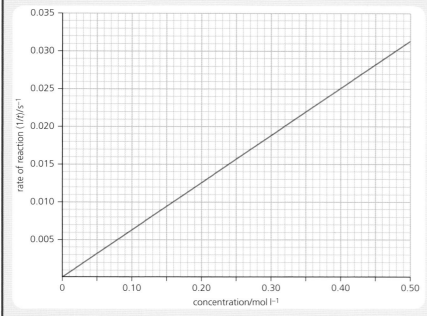

**Figure 1.5**

6   Figure 1.6 shows how the rate of a reaction varies with temperature. The time taken for a colour change to appear was used to measure the rate of the reaction.

a) At what temperature will the colour change take place 200 s after the reagents are mixed?

b) At what time after the reagents are mixed will the colour change take place when the temperature of the solution is 48 °C?

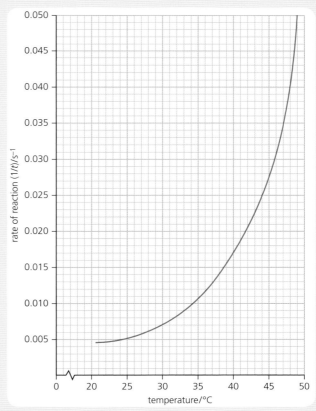

**Figure 1.6**

7   Figure 1.7 shows how the rate of a reaction varies
    with the concentration of one of the reactants. The
    rate is expressed as the reciprocal of the time taken
    for a colour change to appear.

    **a)** How long will the colour change take to appear if
    the reagent concentration is 0.0054 mol l$^{-1}$?

    **b)** At what concentration of reagent will the colour
    change take 40 s to appear?

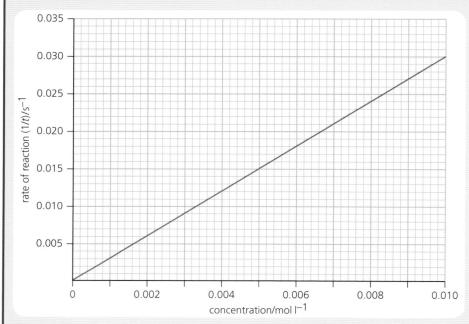

**Figure 1.7**

# 2 Calculations from potential energy graphs

Potential energy diagrams are used in chemistry to illustrate the changes in energy that occur as reactants are converted into products. In this chapter, you will practise interpreting potential energy diagrams so that you can calculate both enthalpy changes and activation energies.

## Enthalpy change

Burning methane gas is an everyday example of a chemical reaction that releases lots of energy.

Figure 2.1 Methane is the gas used in homes when cooking on a gas hob

The actual amount of energy released can be calculated by working out the difference in energy between the reactants and products. This difference is known as the **enthalpy change** and is given the symbol $\Delta H$.

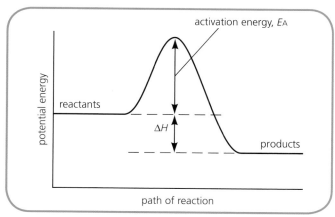

Figure 2.2 A potential energy diagram for an exothermic reaction such as the burning of methane

The potential energy diagram shown in Figure 2.2 could be used to represent the burning of methane.

This diagram shows that the products of the reaction have less energy than the reactants. Where has the energy gone? The difference in energy (the $\Delta H$) will be released to the surroundings resulting in the surroundings rising in temperature. In the case of burning methane gas, the energy released heats up the surrounding air. This released energy can then be used to heat up a pan of water, for example.

## Calculating the enthalpy change

If we know the average energies of the reactants and products, we can calculate the enthalpy change using the equation:

$$\Delta H = H_{products} - H_{reactants}$$

This is explained in the following worked examples.

### Worked example 2.1

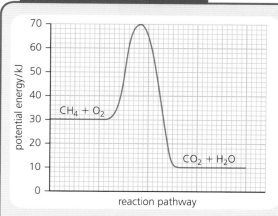

Figure 2.3

**Calculate the enthalpy change for the burning of methane ($CH_4$) as illustrated in the potential energy diagram shown in Figure 2.3.**

**Solution**

The enthalpy change ($\Delta H$) is the difference in energy between the products and reactants.

In this case, $\Delta H = 10 - 30 = -20\,kJ$

## Worked example 2.2

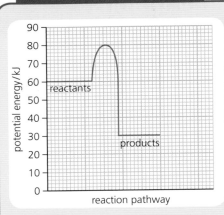

Figure 2.4

**Calculate the enthalpy change for the reaction illustrated in the potential energy diagram shown in Figure 2.4.**

### Solution

The enthalpy change ($\Delta H$) is the difference in energy between the products and reactants.

In this case, $\Delta H = 30 - 60 = \textbf{-30 kJ}$

## Worked example 2.3

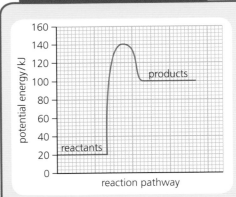

Figure 2.5

**Calculate the enthalpy change for the reaction illustrated in the potential energy diagram shown in Figure 2.5.**

### Solution

The enthalpy change ($\Delta H$) is the difference in energy between the products and reactants.

In this case, $\Delta H = 100 - 20 = \textbf{80 kJ}$

# Activation energy

Methane will only start to burn when ignited by a flame or spark. This is a consequence of the fact that the reaction has a high **activation energy**. This is defined as the minimum energy required for a successful reaction.

Activation energy is represented on potential energy diagrams as shown in Figure 2.6.

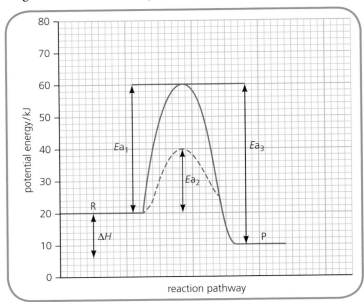

**Figure 2.6** Potential energy diagram illustrating three types of activation energy

This graph shows three types of activation energy:

- $Ea_1$ is the activation energy for the forward reaction i.e. the reaction in which reactants are converted into products.

- $Ea_2$ is the activation energy for the forward reaction when using a **catalyst**. Catalysts work by lowering the activation energy.

- $Ea_3$ is the activation energy for the reverse reaction, i.e. the reaction in which products are converted back into reactants.

The following worked examples show how the activation energy can be calculated from potential energy diagrams.

## Worked example 2.4

Look back at Figure 2.3.

Calculate the activation energy for

a) the forward reaction

b) the reverse reaction.

### Solution

a) The activation energy for the forward reaction is the difference between the peak energy (70 kJ) and the energy of the reactants (30 kJ). In this example, the activation energy is **40 kJ**.

b) The activation energy for the reverse reaction is the difference between the peak energy (70 kJ) and the energy of the products (10 kJ). In this example, the activation energy is **60 kJ**.

## Worked example 2.6

Look back at Figure 2.5.

Calculate the activation energy for

a) the forward reaction

b) the reverse reaction.

### Solution

a) The activation energy for the forward reaction is the difference between the peak energy (140 kJ) and the energy of the reactants (20 kJ). In this example, the activation energy is **120 kJ**.

b) The activation energy for the reverse reaction is the difference between the peak energy (140 kJ) and the energy of the products (100 kJ). In this example, the activation energy is **40 kJ**.

## Worked example 2.5

Look back at Figure 2.4.

Calculate the activation energy for

a) the forward reaction

b) the reverse reaction.

### Solution

a) The activation energy for the forward reaction is the difference between the peak energy (80 kJ) and the energy of the reactants (60 kJ). In this example, the activation energy is **20 kJ**.

b) The activation energy for the reverse reaction is the difference between the peak energy (80 kJ) and the energy of the products (30 kJ). In this example, the activation energy is **50 kJ**.

## Questions

1  For the forward reaction, calculate
   a) the activation energy
   b) the enthalpy change.

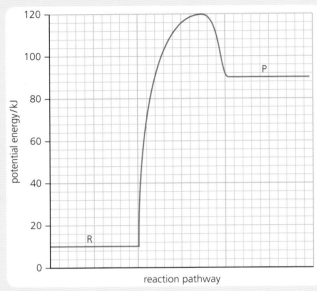

**Figure 2.7**

2  For the forward reaction, calculate
   a) the activation energy
   b) the enthalpy change.

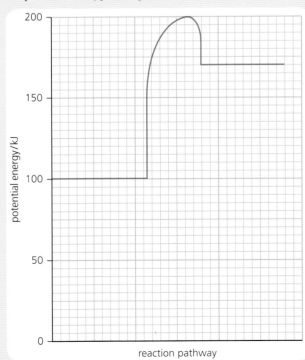

**Figure 2.8**

3  For the reverse reaction, calculate
   a) the activation energy
   b) the enthalpy change.

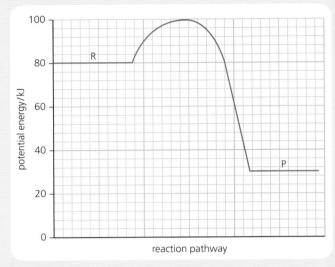

**Figure 2.9**

4  a) For the forward reaction, calculate
      (i) the activation energy
      (ii) the enthalpy change.
   b) Calculate the activation energy for the forward catalysed reaction, represented by the dashed line.

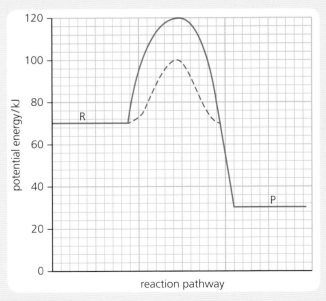

**Figure 2.10**

5  a) Which letter represents the enthalpy change for the forward reaction?

   b) Which letter represents the activation energy for the forward reaction?

   c) Which of the following could be an expression for the activation energy for the reverse reaction?

   A a + b          B a – b

   C a              D b

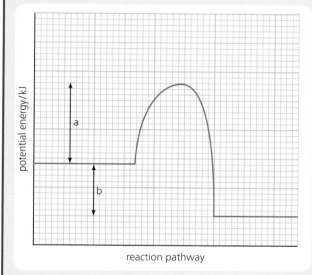

**Figure 2.11**

6  a) Which letter represents the enthalpy change for the forward reaction?

   b) Which letter represents the activation energy for the forward reaction?

   c) Which letter represents the activation energy for the catalysed forward reaction?

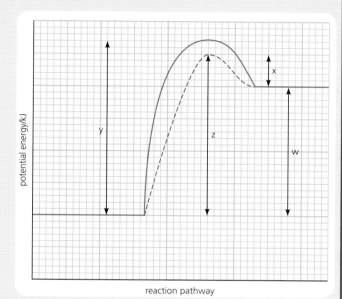

**Figure 2.12**

7  Draw a potential energy diagram using the following information:

   average energy of reactants = 40 kJ

   average energy of products = 100 kJ

   activation energy for the forward reaction = 80 kJ

8  Draw a potential energy diagram using the following information:

   average energy of reactants = 80 kJ

   average energy of products = 10 kJ

   activation energy for the forward reaction = 40 kJ

   activation energy for the forward catalysed reaction = 20 kJ

# Practice with mole calculations

For Higher Chemistry, it is assumed that you are comfortable working with mole calculations that involve:

- using mass and the formula mass
- using concentrations and volumes.

This chapter is designed to help you revise these concepts so that you are prepared for tackling the Higher problems that rely on these relationships.

## The mole as a mass

The mass of 1 mol of any substance is the formula mass in grams. (Note that the accepted abbreviation for the mole is **mol**.)

For example:

1 mol of Cu = 63.5 g

1 mol of $H_2$ = 2 g

1 mol of NaCl = 58.5 g

1 mol of $CO_2$ = 44 g

Mol, mass and formula mass are connected by the equation

$$\text{mass} = \text{mol} \times \text{gfm}$$

where gfm is the formula mass in grams.

Rearranging this equation allows you to calculate moles using

$$\text{mol} = \frac{\text{mass}}{\text{gfm}}$$

If you know the mass of 1 mol, you can work out the mass of different quantities of mol.

For example:

2 mol of $H_2$ = 2 × 2 g = 4 g

100 mol of $CO_2$ = 100 × 44 = 4400 g

Likewise, if you know the mass of a substance, you can work out the number of moles.

For example:

the number of moles present in 88 g of $CO_2$ is $\frac{88}{44}$ = 2 mol

the number of moles present in 72 g of $H_2O$ is $\frac{72}{18}$ = 4 mol

### Worked example 3.1

**Calculate the mass of 2.5 mol of copper.**

**Solution**

mass = mol × gfm

= 2.5 × 63.5

= **158.75 g**

### Worked example 3.2

**How many moles of sodium chloride are present in 11.7 g of the salt?**

**Solution**

mol = $\frac{\text{mass}}{\text{gfm}}$

= $\frac{11.7}{58.5}$

= **0.2 mol**

## Questions

1  Calculate the mass of 0.2 mol of calcium nitrate, $Ca(NO_3)_2$.

2  How many moles of aluminium carbonate, $Al_2(CO_3)_3$, are present in 4.68 g of the substance?

3  How many moles of ammonium carbonate, $(NH_4)_2CO_3$, are present in 115.2 g of the substance?

4  Calculate the mass of 0.025 mol of sucrose, $C_{12}H_{22}O_{11}$.

5  Washing soda, sodium carbonate-10-water, has the formula $Na_2CO_3.10H_2O$. How many moles of washing soda are present in 0.715 g of the substance?

6  Calculate the mass of 0.4 mol of ammonium phosphate, $(NH_4)_3PO_4$.

7  How many moles of water, $H_2O$, are present in 5.4 g of the substance?

8  Calculate the mass of 0.08 mol of carbon monoxide, CO.

9  How many moles of barium chloride, $BaCl_2$, are present in 4.166 g of the salt?

10  Calculate the mass of 1.2 mol of sodium hydroxide, NaOH.

11  Calculate the mass of 0.2 mol of copper(II) chloride, $CuCl_2$.

12  How many moles of nitric acid, $HNO_3$, are present in 94.5 g of the pure substance?

13  Calculate the mass of 0.025 mol of iron(III) oxide, $Fe_2O_3$.

14  How many moles of silver(I) nitrate, $AgNO_3$, are present in 6.796 g of the substance?

15  Calculate the mass of 3.5 mol of propene, $C_3H_6$.

16  How many moles of ammonium sulfate, $(NH_4)_2SO_4$, are present in 105.68 g of the substance?

17  Calculate the mass of 0.5 mol of copper(II) sulfate-5-water, $CuSO_4.5H_2O$.

18  How many moles of mercury(II) nitrate, $Hg(NO_3)_2$, are present in 16.23 g of the substance?

19  Calculate the mass of 0.3 mol of sodium carbonate, $Na_2CO_3$.

20  How many moles of aluminium hydrogensulfite, $Al(HSO_3)_3$, are present in 8.109 g of the substance?

# The mole, volume and concentration

In chemical terms, the concentration of a solution is the number of moles of substance dissolved in each litre of the solution. The unit for this is '**moles per litre**' which is abbreviated as **mol l⁻¹**.

Although not used in this book, or in SQA examinations, you will sometimes see units of concentration shown as 'M' or 'mol/l'. Both abbreviations mean the same thing (mole per litre) but we will stick to the correct abbreviation which is mol l⁻¹.

The relationship between moles, concentration and volume is shown by the expression:

$$\text{concentration} = \frac{\text{number of moles}}{\text{volume in litres}}$$

This is often shown in the form of a triangle, as illustrated in Figure 3.1.

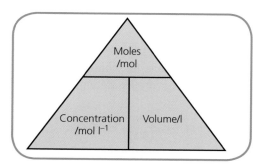

**Figure 3.1** The relationship between concentration, number of moles and volume in litres

The other relationships can be expressed as:

$$\text{volume (in litres)} = \frac{\text{number of moles}}{\text{concentration}}$$

$$\text{number of moles} = \text{concentration} \times \text{volume (in litres)}$$

## Worked example 3.3

**What is the concentration of a solution containing 2.5 mol of substance dissolved in 5 l of solution?**

### Solution

$$\text{concentration} = \frac{\text{number of moles}}{\text{volume in litres}}$$

$$= \frac{2.5}{5}$$

$$= 0.5 \, mol \, l^{-1}$$

## Worked example 3.4

**How many moles of substance are present in 25 cm$^3$ of a 0.2 mol l$^{-1}$ solution?**

### Solution

In this problem, the volume of solution has been given in cm$^3$ and must be converted to litres.

In most laboratory chemistry, only small volumes of solutions are used in practice, so the smaller unit of the **cubic centimetre**, abbreviated to cm$^3$, is commonly used.

$$1 \, litre = 1000 \, cm^3$$

$$1 \, cm^3 = \frac{1}{1000} \, litre$$

$$= 0.001 \, litre$$

Note that the term **millilitre**, abbreviated to **ml**, is often used as a unit of volume for measuring household liquid volumes, e.g. drinks, detergents etc. This unit is identical to cm$^3$.

Since 1 l = 1000 cm$^3$, 25 cm$^3$ is $\frac{25}{1000} l = 0.025 l$.

This value can now be fitted into the appropriate equation:

$$\text{number of moles} = \text{concentration} \times \text{volume (in litres)}$$

$$= 0.2 \times 0.025$$

$$= 0.005 \, mol$$

## Worked example 3.5

2 g of sodium hydroxide, NaOH, are dissolved in water to make a 0.4 mol l$^{-1}$ solution. What volume is the solution?

In order to calculate the volume we need to know the concentration and the number of moles, but we are only told the concentration; we must first work out the number of moles of sodium hydroxide from its mass:

formula of sodium hydroxide: NaOH

formula mass: 23 + 16 + 1 = 40

$$\text{number of moles} = \frac{2}{40} = 0.05 \, mol$$

We now fit this value, and that of the concentration, into the appropriate equation:

$$\text{volume (in litres)} = \frac{\text{number of moles}}{\text{concentration}}$$

$$= \frac{0.05}{0.4}$$

$$= 0.125 \, l \, (125 \, cm^3)$$

## Worked example 3.6

**What mass of sodium carbonate, Na$_2$CO$_3$, must be dissolved to make 0.25 litres of a 0.2 mol l$^{-1}$ solution?**

### Solution

$$\text{number of moles} = \text{concentration} \times \text{volume (in litres)}$$

$$= 0.2 \times 0.25$$

$$= 0.05 \, mol$$

However, the question asks for the **mass** of sodium carbonate which this represents.

formula of sodium carbonate: Na$_2$CO$_3$

formula mass: $(2 \times 23) + 12 + (3 \times 16) = 106$

1 mol of sodium carbonate = 106 g

0.05 mol of sodium carbonate = 0.05 × 106 g

$$= 5.3 \, g$$

## Questions

**Questions 21 to 25 are of the type shown in worked examples 3.3 and 3.4. Questions 26 to 40 are of the type shown in worked examples 3.5 and 3.6.**

21  0.24 mol of salt is dissolved to make 1.2 l of solution. What is the concentration of the solution?

22  200 cm³ of a salt solution has a concentration of 1.5 mol l⁻¹. How many moles of salt are dissolved in it?

23  0.005 mol of a substance is dissolved in 25 cm³ of solution. What is the concentration of the solution?

24  What is the volume (in cm³) of a 1.2 mol l⁻¹ solution which contains 0.048 mol of dissolved substance?

25  A flask contains 300 cm³ of a 0.5 mol l⁻¹ acid solution. How many moles of pure acid must have been dissolved?

26  2.943 g of pure sulfuric acid, $H_2SO_4$, is dissolved in water to make 150 cm³ of solution. What is the concentration of the acid solution now?

27  A 0.2 mol l⁻¹ solution of sodium carbonate, $Na_2CO_3$, is made by dissolving 5.3 g of the solid in water and making it up to the mark in a standard flask. What volume must the flask be?

28  0.8 l of a 0.5 mol l⁻¹ solution of ammonium nitrate, $NH_4NO_3$, has to be made up. What mass of solid ammonium nitrate would be required?

29  71.05 g of sodium sulfate, $Na_2SO_4$, is dissolved to make a 2 l standard solution. What concentration is the solution?

30  25 cm³ of a 0.4 mol l⁻¹ solution of ammonium sulfate, $(NH_4)_2SO_4$, is made up. What mass of solid must have been dissolved?

31  A 0.2 mol l⁻¹ solution of potassium nitrate, $KNO_3$, is made by dissolving 30.33 g of the solid in a standard flask. What volume is the standard flask?

32  0.4 g of sodium hydroxide, NaOH, is dissolved in water to make a 0.25 mol l⁻¹ solution. What volume is the solution?

33  2.675 g of ammonium chloride, $NH_4Cl$, is dissolved in water to make a 0.1 mol l⁻¹ solution. What volume of solution is made?

34  What mass of anhydrous copper(II) sulfate, $CuSO_4$, would be required to make 100 cm³ of a 0.5 mol l⁻¹ solution?

35  What would be the concentration of a 200 cm³ solution of silver(I) nitrate, $AgNO_3$, containing 1.699 g of dissolved solid?

36  1.92 g of ammonium carbonate, $(NH_4)_2CO_3$, is dissolved to make 400 cm³ of solution. What is the concentration of the solution?

37  2.764 g of potassium carbonate, $K_2CO_3$, is dissolved in 200 cm³ of solution. What is the concentration of the solution?

38  What mass of pure ethanoic acid, $CH_3COOH$, would be required to make 40 cm³ of a 0.4 mol l⁻¹ solution?

39  14.3 g of sodium carbonate-10-water, $Na_2CO_3.10H_2O$, is dissolved in water and made up to 250 cm³ in a standard flask. What is the concentration of the sodium carbonate solution?

40  Oxalic acid has the formula $(COOH)_2$. If 0.0225 g of the pure acid was dissolved in water to make a 0.01 mol l⁻¹ solution, what would be the volume?

# 4 Calculations from equations: revision of previous work

Calculations from equations allow chemists to work out how much of a product can be made from a certain quantity of reactant. Alternatively, they can be used to allow a chemist to calculate how much reactant to use in order to produce a certain quantity of product.

This chapter allows you to practise with simple calculations that involve one reactant and should help you revise work you have already done. Once you have mastered this technique, you can try the more demanding calculations in Chapter 5 which require you to work with the quantities of two reactants.

## Understanding how to use balanced chemical equations

When methane gas is burned, carbon dioxide and water are produced. We can express this using a balanced chemical equation:

$$CH_4 + 2O_2 \rightarrow CO_2 + 2H_2O$$

This equation tells us that one methane molecule reacts with two oxygen molecules to produce one carbon dioxide molecule and two water molecules. In reality, chemists do not work with such small quantities (one or two molecules). Instead, they work with the **mole** (shortened to mol). The mole is simply a quantity in the same way that one dozen is a quantity. If we apply this to a balanced chemical equation, we would say that:

1 mol of methane reacts with 2 mol of oxygen to produce 1 mol of carbon dioxide and 2 mol of water.

Since we know the relationship between mole and mass, we can convert the mole to mass and *vice versa*. Let's look at some examples.

## Calculations from equations: quantities in grams

### Worked example 4.1

Calculate the mass of carbon dioxide produced when 640 g of methane gas is exploded in excess oxygen according to the following equation:

$$CH_4(g) + 2O_2(g) \rightarrow CO_2(g) + 2H_2O(l)$$

**Solution**

**Step 1: Calculate the number of moles of the reagent you are given a mass for**

In this case, the only reagent we can consider is methane since this is the only reagent we have a mass for.

$$moles = \frac{mass}{gfm} = \frac{640}{16} = 40\,mol$$

**Step 2: Use the mole ratio to calculate the number of moles of product**

From the equation, 1 mol of methane produces 1 mol of carbon dioxide.

So, 40 mol $CH_4 \rightarrow$ 40 mol $CO_2$

**Step 3: Convert moles to mass**

mass = moles × gfm

= 40 × 44 (where 44 is the gfm of $CO_2$)

= **1760 g**

A common question at this final step is: 'Which gfm should I use?'

You should always use the gfm of the reactant or product you are trying to calculate. In this case, you have calculated that 40 mol of $CO_2$ will be produced. You are now trying to convert this into a mass of $CO_2$. So, when you write the equation, mass = mol × gfm, it is the gfm of $CO_2$ you use (44) since this is the mass you are trying to calculate.

## Worked example 4.2

What mass of copper would be obtained by heating 3.975g of copper(II) oxide with an excess of carbon?

$$2CuO + C \rightarrow 2Cu + CO_2$$

### Solution

**Step 1: Calculate the number of moles of the reagent you are given a mass for**

In this case, the only reagent we can consider is $CuO$ since this is the only reagent we have a mass for:

$$moles = \frac{mass}{gfm}$$

$$= \frac{3.975}{79.5}$$

$$= 0.05\,mol$$

(where 79.5 is the gfm of $CuO$)

**Step 2: Use the mole ratio to calculate the number of moles of product**

From the equation, 2mol of $CuO$ produces 2mol of $Cu$. In other words, 1mol $CuO \rightarrow$ 1mol $Cu$.

So, 0.05mol $CuO \rightarrow$ 0.05mol $Cu$

**Step 3: Convert moles to mass**

$$mass = moles \times gfm$$

$$= 0.05 \times 63.5$$

(where 63.5 is the gfm of $Cu$)

$$= 3.175\,g$$

Note that in Step 1 we use the gfm for $CuO$ since we are trying to calculate the number of moles of $CuO$ heated. In Step 3, however, we use the gfm of $Cu$ since we are trying to calculate the mass of $Cu$ produced.

## Worked example 4.3

Calculate the mass of hydrogen required to react completely with excess oxygen to produce 720g of water.

$$H_2 + \frac{1}{2}O_2 \rightarrow H_2O$$

### Solution

**Step 1: Calculate the number of moles of the reagent you are given a mass for**

In this case, the only reagent we can consider is $H_2O$ since this is the only reagent we have a mass for.

$$moles = \frac{mass}{gfm}$$

$$= \frac{720}{18}$$

$$= 40\,mol$$

(where 18 is the gfm of $H_2O$)

**Step 2: Use the mole ratio to calculate the number of moles of reactant**

From the equation, 1mol of water came from 1mol of hydrogen.

So, 40mol $H_2O \rightarrow$ 40mol $H_2$

**Step 3: Convert moles to mass**

$$mass = moles \times gfm$$

$$= 40 \times 2$$

(where 2 is the gfm of $H_2$)

$$= 80\,g$$

## Questions

1   What mass of hydrogen gas would be formed if 6.075 g of magnesium reacted completely with dilute sulfuric acid?

$Mg + H_2SO_4 \rightarrow MgSO_4 + H_2$

2   What mass of copper metal would be formed by the complete reaction of 15.9 g of copper(II) oxide with hydrogen gas?

$CuO + H_2 \rightarrow Cu + H_2O$

3   Calculate the mass of carbon required to react completely with 22.32 g of lead(II) oxide.

$2PbO + C \rightarrow 2Pb + CO_2$

4   Calculate the mass of carbon monoxide required to form 5.5 g of carbon dioxide.

$2CO + O_2 \rightarrow 2CO_2$

5   0.5 g of hydrogen gas is given off when sodium is reacted completely with water. Calculate the mass of sodium that reacted.

$2Na + 2H_2O \rightarrow 2NaOH + H_2$

6   13.08 g of zinc is reacted with excess hydrochloric acid. Calculate the mass of hydrogen produced.

$Zn + 2HCl \rightarrow ZnCl_2 + H_2$

7   What mass of magnesium will react completely with 36 g of pure ethanoic acid, $CH_3COOH$?

$Mg + 2CH_3COOH \rightarrow Mg(CH_3COO)_2 + H_2$

8   5.3 g of sodium carbonate is reacted with an excess of sulfuric acid. What mass of carbon dioxide would be evolved?

$Na_2CO_3 + H_2SO_4 \rightarrow Na_2SO_4 + CO_2 + H_2O$

9   45 g of glucose is burned completely in an excess of oxygen. Calculate the mass of carbon dioxide produced.

$C_6H_{12}O_6 + 6O_2 \rightarrow 6CO_2 + 6H_2O$

10  What mass of oxygen would be produced by the complete reaction of 27.58 g of silver(I) carbonate?

$2Ag_2CO_3 \rightarrow 4Ag + 2CO_2 + O_2$

# Calculations from equations: quantities in kilograms or tonnes

## Worked example 4.4

Hydrazine, $N_2H_4$, is a rocket fuel which reacts with oxygen producing gaseous products of nitrogen and water vapour under very high pressures.

$$N_2H_4 + O_2 \rightarrow N_2 + 2H_2O$$

If $6.4 \times 10^4$ kg of hydrazine is burned completely, what mass of water vapour will be produced?

### Solution

The only difference between this question and the previous type is that the quantity referred to is expressed as a large number of kilograms, rather than a small number of grams. This refers to a process being carried out on an industrial scale instead of using laboratory-size quantities. Another unit of mass which can be involved in industrial quantities is the **tonne** which is equal to 1000 kg.

The easiest way to approach this is to work out the answer as if the question had referred to 6.4 **grams** of hydrazine; the conversion to **kilograms** can wait until the final step of the problem, as shown below.

### Step 1: Calculate the number of moles of the reagent you are given a mass for

In this case, the only reagent we can consider is hydrazine since this is the only reagent we have a mass for.

$$moles = \frac{mass}{gfm}$$

$$= \frac{6.4}{32}$$

$$= 0.2 \, mol$$

(where 32 is the gfm of $N_2H_4$)

### Step 2: Use the mole ratio to calculate the number of moles of product

From the equation, 1 mol of hydrazine reacts to form 2 mol of water.

So, $0.2 \, mol \, N_2H_4 \rightarrow 0.4 \, mol \, H_2O$

### Step 3: Convert moles to mass

$$mass = moles \times gfm$$

$$= 0.4 \times 18$$

(where 18 is the gfm of $H_2O$)

$$= 7.2 \, g$$

This would be the end of the question if it had asked us about the reaction of 6.4 g of $N_2H_4$; however, the question referred to $6.4 \times 10^4$ kg of $N_2H_4$. A simple bit of direct proportion gives us the answer:

6.4 g of $N_2H_4$ reacts to form 7.2 g of $H_2O$

**$6.4 \times 10^4$ kg of $N_2H_4$ reacts to form $7.2 \times 10^4$ kg of $H_2O$**

## Questions

11  Assuming 100% conversion of reactants to products, what mass of sulfur dioxide would be required to produce 1602 kg of sulfur trioxide by the Contact Process described by the following equation?

$$2SO_2 + O_2 \rightarrow 2SO_3$$

12  175 kg of nitrogen is converted completely to ammonia in the Haber Process represented by the following equation. What mass of hydrogen must have reacted?

$$N_2 + 3H_2 \rightarrow 2NH_3$$

13  63.84 tonnes of iron(III) oxide is reduced completely by carbon monoxide to form pure iron. Calculate the mass of iron produced.

$$Fe_2O_3 + 3CO \rightarrow 2Fe + 3CO_2$$

14  7692 kg of sulfur dioxide is converted to sulfur trioxide by the following process. Calculate the mass of sulfur trioxide produced.

$$2SO_2 + O_2 \rightarrow 2SO_3$$

15  Ethanol, $C_2H_5OH$, can be produced industrially by the catalytic reaction of ethene with steam. What mass of ethene would be needed to produce 1104 kg of ethanol?

$$C_2H_4 + H_2O \rightarrow C_2H_5OH$$

16  The final stage in the industrial production of nitric acid, $HNO_3$, involves the following reaction. What mass of nitrogen dioxide must react to produce $2.52 \times 10^3$ kg of nitric acid?

$$3NO_2 + H_2O \rightarrow 2HNO_3 + NO$$

17  The following equation shows the catalytic hydration of ethyne, $C_2H_2$, to ethanal, $CH_3CHO$. If $2.08 \times 10^4$ kg of ethyne is reacted completely, calculate the mass of ethanal produced.

$$C_2H_2 + H_2O \rightarrow CH_3CHO$$

18  Titanium metal can be extracted from titanium chloride using displacement by sodium. What mass of sodium would be needed to react completely with 7.596 tonnes of titanium chloride?

$$TiCl_4 + 4Na \rightarrow Ti + 4NaCl$$

19  Calculate the mass of fluorine required to react completely with $8 \times 10^4$ kg of hydrazine, $N_2H_4$.

$$N_2H_4 + 2F_2 \rightarrow N_2 + 4HF$$

20  Calculate the mass of ethanal, $CH_3CHO$, required to produce $2.95 \times 10^4$ kg of trichloroethanal, $CCl_3CHO$.

$$CH_3CHO + 3Cl_2 \rightarrow CCl_3CHO + 3HCl$$

# 5 Calculations from equations: excess reagent

In Chapter 4, a quantity of one substance was given and the quantity of another was asked for. In this chapter, the quantities of two reactants will be given and the quantity of a product will be asked for. These calculations are known as excess calculations since there will always be a surplus of one of the reactants.

## Understanding the meaning of 'in excess' and 'limiting reagent'

Previous examples have considered the effect of one reactant. For example, calculate the mass of hydrogen chloride formed when 40 g of hydrogen reacts with excess chlorine in the following reaction:

$$H_2 + Cl_2 \rightarrow 2HCl$$

To tackle this example, you only have to consider one reactant. We ignore the effect of the quantity of chlorine because we are told it is in excess, i.e. there is a plentiful supply of chlorine to ensure that all the hydrogen is reacted. The quantity of hydrogen chloride produced depends on the quantity of hydrogen. Hydrogen is said to be the **limiting reagent**.

In this chapter, we will consider examples where data are given for *two* reagents. For example, calculate the mass of hydrogen chloride formed when 40 g of hydrogen reacts with 710 g of chlorine in the following reaction:

$$H_2 + Cl_2 \rightarrow 2HCl$$

In these examples, one of the reactants is **in excess**; in other words, there is too much of it to react with all of the other reactant.

To calculate the amount of product made, we need to calculate the number of moles of both reactants and identify which one is in excess. We then use the number of moles of the other reactant (the limiting reagent) as our 'known' moles, since we know that *all* of it will react.

## Calculations from equations: straightforward examples

### Worked example 5.1

40 g of hydrogen was reacted with 710 g of chlorine:

$$H_2 + Cl_2 \rightarrow 2HCl$$

a) Calculate which reactant was in excess.

b) Calculate the mass of hydrogen chloride formed.

**Solution**

**Step 1: Calculate the number of moles of each reactant**

moles of hydrogen $= \frac{40}{2} = 20$ mol

moles of chlorine $= \frac{710}{71} = 10$ mol

**Step 2: Calculate which reactant is in excess by looking at the mole ratio**

According to the equation, 1 mol of $H_2$ reacts with 1 mol of $Cl_2$.

So, 20 mol of hydrogen would require 20 mol of chlorine for complete reaction.

Since there is only 10 mol of chlorine, we do not have enough to react completely with the hydrogen. In other words, the hydrogen is in excess.

**Step 3: Calculate the moles, and hence mass, of product formed**

$$H_2 + Cl_2 \rightarrow 2HCl$$

We are now focusing on the limiting reagent (chlorine) since the hydrogen is in excess.

1 mol of chlorine $\rightarrow$ 2 mol of hydrogen chloride

10 mol of chlorine $\rightarrow$ 20 mol of hydrogen chloride

mass = moles × gfm

     = 20 × 36.5

     = **730 g**

## Worked example 5.2

**18 g of methane gas was reacted with 24 g of oxygen.**

$$CH_4(g) + 2O_2(g) \rightarrow CO_2(g) + 2H_2O(g)$$

a) Calculate which reactant is in excess.

b) Calculate the mass of carbon dioxide gas produced.

### Solution

**Step 1: Calculate the number of moles of each reactant**

moles of methane = $\frac{18}{16}$ = 1.125 mol

moles of oxygen = $\frac{24}{32}$ = 0.75 mol

**Step 2: Calculate which reactant is in excess by looking at the mole ratio**

According to the equation, 1 mol of $CH_4$ reacts with 2 mol of $O_2$.

So, 1.125 mol of methane would require 2.25 mol of oxygen for complete reaction.

Since there is only 0.75 mol of oxygen, we do not have enough to react completely with the methane. In other words, the methane is in excess.

**Step 3: Calculate the moles, and hence mass, of product formed**

$$CH_4(g) + 2O_2(g) \rightarrow CO_2(g) + 2H_2O(g)$$

We are now focusing on the limiting reagent (oxygen) since the methane is in excess.

2 mol oxygen → 1 mol carbon dioxide

0.75 mol oxygen → 0.375 mol carbon dioxide

mass = moles × gfm

= 0.375 × 44

= 16.5 g

## Questions

1 24 g of carbon was reacted with 38 g of oxygen.

$$C(s) + \frac{1}{2}O_2(g) \rightarrow CO(g)$$

   a) Calculate which reactant is in excess.

   b) Calculate the mass of carbon monoxide formed.

2 78 g of methane was exploded with 18 g of oxygen.

$$CH_4(g) + 2O_2(g) \rightarrow CO_2(g) + 2H_2O(l)$$

   a) Calculate which reactant is in excess.

   b) Calculate the mass of carbon dioxide formed.

3 6.216 g of lead is added to 50 cm³ of 1 mol l⁻¹ hydrochloric acid.

$$Pb(s) + 2HCl(aq) \rightarrow PbCl_2(aq) + H_2(g)$$

   a) Calculate which reactant is in excess.

   b) Calculate the mass of hydrogen formed.

4 12 g of calcium carbonate is reacted with 500 cm³ of 0.4 mol l⁻¹ nitric acid.

$$CaCO_3(s) + 2HNO_3(aq) \rightarrow Ca(NO_3)_2(aq) + CO_2(g) + H_2O(l)$$

   a) Calculate which reactant is in excess.

   b) Calculate the mass of carbon dioxide produced.

5 120 cm³ of 0.2 mol l⁻¹ lead(II) nitrate solution is added to 200 cm³ of 0.25 mol l⁻¹ potassium iodide solution. The lead(II) iodide precipitate formed is filtered and dried.

$$Pb(NO_3)_2(aq) + 2KI(aq) \rightarrow PbI_2(s) + 2KNO_3(aq)$$

   a) Calculate which reactant is in excess.

   b) What is the theoretical mass of precipitate obtained?

# Calculations from equations: more challenging examples

## Challenge 1: recognising that it is an excess calculation

The questions in the previous examples made it clear that you had to apply the excess calculation method as this was stated in the question, 'Calculate which reactant is **in excess**'. The examples which follow are more challenging as it is not always immediately obvious that you are dealing with an excess calculation; it won't be stated in the question. However, it is easy to identify an excess question: you are always given information about **two** of the reactants.

---

### Worked example 5.3

1.308 g of zinc is reacted with 25 cm³ of 2 mol l⁻¹ hydrochloric acid.

$$Zn(s) + 2HCl(aq) \rightarrow ZnCl_2(aq) + H_2(g)$$

**Calculate the mass of hydrogen gas formed.**

#### Solution

In this case you need to recognise that this is an excess calculation. The clue is that you are given information about **two** reactants.

**Step 1: Calculate the number of moles of each reactant**

$$\text{moles of zinc} = \frac{1.308}{65.4}$$

$$= 0.02 \text{ mol}$$

$$\text{moles of hydrochloric acid} = CV$$

$$= 2 \times \frac{25}{1000}$$

$$= 0.05 \text{ mol}$$

**Step 2: Calculate which reactant is in excess by looking at the mole ratio**

According to the equation, 1 mol of Zn reacts with 2 mol of HCl.

So, 0.02 mol of zinc would require 0.04 mol of hydrochloric acid for complete reaction.

Since we have 0.05 mol of hydrochloric acid, the hydrochloric acid is in excess.

**Step 3: Calculate the moles, and hence mass, of product formed**

$$Zn(s) + 2HCl(aq) \rightarrow ZnCl_2(aq) + H_2(g)$$

We are now focusing on the limiting reagent (zinc) since the acid is in excess.

1 mol of zinc → 1 mol of hydrogen

0.02 mol of Zn → 0.02 mol of hydrogen

$$\text{mass} = \text{moles} \times \text{gfm}$$

$$= 0.02 \times 2$$

$$= 0.04 \text{ g}$$

## Challenge 2: working with a complicated mole ratio

Another challenge can be dealing with calculations where the mole ratio is not straightforward. The previous examples have used quite simple mole ratios of 1:1 or 1:2. The following worked example has a more complicated mole ratio. The method used to tackle this is the same as we have used to tackle previous examples.

### Worked example 5.4

**Calculate the mass of water formed when 60.2 g of hexane was reacted with 160 g of oxygen gas.**

$$C_6H_{14}(g) + 9.5O_2(g) \rightarrow 6CO_2(g) + 7H_2O(g)$$

**Solution**

**Step 1: Calculate the number of moles of each reactant**

$$\text{moles of hexane} = \frac{60.2}{86}$$
$$= 0.7 \, \text{mol}$$

$$\text{moles of oxygen} = \frac{160}{32}$$
$$= 5 \, \text{mol}$$

**Step 2: Calculate which reactant is in excess by looking at the mole ratio**

According to the equation, 1 mol of hexane reacts with 9.5 mol of oxygen.

So, 0.7 mol of hexane would require

$9.5 \times 0.7 = 6.65$ mol of oxygen.

Since there is only 5 mol of oxygen (i.e. less than the required 6.65), the hexane is in excess.

**Step 3: Calculate the moles, and hence mass, of product formed**

$$C_6H_{14}(g) + 9.5O_2(g) \rightarrow 6CO_2(g) + 7H_2O(g)$$

We are now focusing on the limiting reagent (oxygen) since the hexane is in excess.

9.5 mol of oxygen $\rightarrow$ 7 mol of water

1 mol of oxygen $\rightarrow \dfrac{7}{9.5} = 0.737$ mol of water

5 mol of oxygen $\rightarrow 0.737 \times 5 = 3.68$ mol of water

$$\text{mass} = \text{moles} \times \text{gfm}$$
$$= 3.68 \times 18$$
$$= \mathbf{66.3 \, g}$$

The trick in the final part of this question was to use proportion to work out how many moles of water would be produced for 1 mol of oxygen. This was achieved by dividing by 9.5; since there are 9.5 mol of oxygen; dividing by 9.5 ensures that we have 1 mol of oxygen.

## Challenge 3: demonstrating that a reactant is in excess

Another common approach in excess questions is to ask you to 'Show by calculation that reactant X is in excess'. This is the same as asking you to calculate the excess reactant, except that you must show how you know this. The steps are exactly the same. This is shown in the following worked example.

### Worked example 5.5

28 g of iron was reacted with 200 cm³ of 0.2 mol l⁻¹ sulfuric acid.

$$Fe(s) + H_2SO_4(aq) \rightarrow FeSO_4(aq) + H_2(g)$$

**Show by calculation that the iron was in excess.**

**Solution**

**Step 1: Calculate the number of moles of each reactant**

moles of iron $= \frac{28}{56}$

$\qquad = 0.5\,mol$

moles of sulfuric acid $= CV$

$\qquad = 0.2 \times \frac{200}{1000}$

$\qquad = 0.04\,mol$

**Step 2: Calculate which reactant is in excess by looking at the mole ratio**

1 mol of iron reacts with 1 mol of sulfuric acid.

0.5 mol of iron will react with 0.5 mol of sulfuric acid. However, we only have 0.04 mol of sulfuric acid so the iron is in excess.

## Challenge 4: calculating the left-over reactant

A final point to note is that some excess questions are extended by asking you to calculate the mass of reactant that is unreacted (or left over). In other words, if you have two reactants and one of them is in excess, there will be some left over. The calculations from worked example 5.3 will be used to illustrate how this works.

### Worked example 5.6

1.308 g of zinc is reacted with 25 cm³ of 2 mol l⁻¹ hydrochloric acid.

$$Zn(s) + 2HCl(aq) \rightarrow ZnCl_2(aq) + H_2(g)$$

**Calculate the number of moles of unreacted hydrochloric acid.**

**Solution**

**Step 1: Calculate the number of moles of each reactant**

moles of zinc $= \frac{1.308}{65.4}$

$\qquad = 0.02\,mol$

moles of hydrochloric acid $= CV$

$\qquad = 2 \times \frac{25}{1000}$

$\qquad = 0.05\,mol$

**Step 2: Calculate which reactant is in excess by looking at the mole ratio**

According to the equation, 1 mol of Zn reacts with 2 mol of HCl.

So, 0.02 mol of zinc would require 0.04 mol of hydrochloric acid for complete reaction.

Since there is 0.05 mol of hydrochloric acid, the hydrochloric acid is in excess.

**Step 3: Calculate how much of the excess reactant is unreacted**

From the previous step, only 0.04 mol of hydrochloric acid is reacted. Therefore, the number of moles of unreacted acid = 0.05 mol – 0.04 mol = 0.01 mol.

## Questions

**Questions 6–9 make use of the concepts discussed in worked examples 5.3 and 5.4.**

6 Calculate the mass of carbon dioxide produced when 32 g of butane is reacted with 18 g of oxygen gas.

$$C_4H_{10}(g) + 6\tfrac{1}{2}O_2(g) \rightarrow 4CO_2(g) + 5H_2O(g)$$

7 Calculate the mass of water formed when 2.8 g of hydrogen is reacted with 18 g of oxygen.

$$2H_2(g) + O_2(g) \rightarrow 2H_2O(g)$$

8 3 g of pure ethanoic acid, $CH_3COOH$, is added to 90 cm³ of 0.4 mol l⁻¹ sodium carbonate solution. Calculate the mass of carbon dioxide produced.

$$2CH_3COOH(l) + Na_2CO_3(aq) \rightarrow 2CH_3COONa(aq) + CO_2(g) + H_2O(l)$$

9 40 g of magnesium was reacted with 200 cm³ of 2 mol l⁻¹ hydrochloric acid. Calculate the mass of hydrogen gas produced.

$$Mg(s) + 2HCl(aq) \rightarrow MgCl_2(aq) + H_2(g)$$

**Questions 10–15 make use of the concepts discussed in worked examples 5.5 and 5.6.**

10 40 cm³ of 0.5 mol l⁻¹ aluminium nitrate solution is added to 200 cm³ of 0.4 mol l⁻¹ sodium hydroxide solution.

$$Al(NO_3)_3(aq) + 3NaOH(aq) \rightarrow Al(OH)_3(s) + 3NaNO_3(aq)$$

a) Show by calculation that the sodium hydroxide was in excess.

b) Calculate the mass of aluminium hydroxide produced.

c) Calculate the number of moles of unreacted sodium hydroxide.

11 40 cm³ of 0.045 mol l⁻¹ silver(I) nitrate solution is added to 12 cm³ of 0.1 mol l⁻¹ magnesium chloride solution.

$$2AgNO_3(aq) + MgCl_2(aq) \rightarrow 2AgCl(s) + Mg(NO_3)_2(aq)$$

a) Show by calculation that the magnesium chloride was in excess.

b) Calculate the mass of silver(I) chloride produced.

c) Calculate the number of moles of unreacted magnesium chloride.

12 20 cm³ of 0.2 mol l⁻¹ sulfuric acid is added to 50 cm³ of 0.1 mol l⁻¹ barium chloride solution.

$$H_2SO_4(aq) + BaCl_2(aq) \rightarrow BaSO_4(s) + 2HCl(aq)$$

a) Show by calculation that the barium chloride was in excess.

b) Calculate the mass of barium sulfate produced.

13 6.605 g of ammonium sulfate is added to 120 cm³ of 0.5 mol l⁻¹ sodium hydroxide solution and the solution heated.

$$(NH_4)_2SO_4(aq) + 2NaOH(aq) \rightarrow Na_2SO_4(aq) + 2H_2O(l) + 2NH_3(g)$$

a) Show by calculation that the ammonium sulfate is in excess.

b) Calculate the mass of ammonia, $NH_3$, produced.

14 4.395 g of iron(II) sulfide is added to 160 cm³ of 0.5 mol l⁻¹ hydrochloric acid.

$$FeS(s) + 2HCl(aq) \rightarrow FeCl_2(aq) + H_2S(g)$$

a) Show by calculation that the iron(II) sulfide is in excess.

b) Calculate the mass of hydrogen sulfide gas formed.

15 1.59 g of copper(II) oxide is added to a beaker of 50 cm³ of 0.24 mol l⁻¹ sulfuric acid. The mixture is heated and stirred until no further reaction takes place, and the contents of the beaker are filtered.

$$CuO(s) + H_2SO_4(aq) \rightarrow CuSO_4(aq) + H_2O(l)$$

a) Work out which reactant is in excess.

b) What mass of unreacted copper(II) oxide would be removed from the beaker by the filtration, after being dried?

c) What mass of copper(II) sulfate would be obtained?

# 6 Molar volume

The calculations we have dealt with thus far have involved solids, liquids or solutions. For these substances, we have expressed the quantities in mass (for solids) or we have used volumes and concentrations for liquids and solutions. In this chapter, we will solve problems involving gases.

## Molar volume

When dealing with gases, we need to consider a new quantity: **molar volume**.

**Molar volume is the volume occupied by 1 mole of gas.**

Experimentally, it is found that the volume occupied by 1 mole of any gas is approximately the same provided the measurements are taken at the same temperature and pressure.

For example, the volume occupied by 1 mole of hydrogen gas at room temperature and 1 atmosphere of pressure is approximately 24 litres. Knowing this, we can state that the volume occupied by 1 mole of methane gas (or any other gas) would also be 24 litres. This fact allows us to solve calculations where the reactant(s) or product(s) are gases.

## Using the molar volume to calculate the number of moles or volume of a gas

Since the molar volume is the volume occupied by 1 mole of a gas, you can use this relationship to calculate the number of moles or volume of a gas. Conversely, you can use the number of moles, or volume of a gas, to calculate the molar volume. The worked examples illustrate how to tackle such questions involving molar volume.

---

### Worked example 6.1

**Calculate the volume occupied by 3 mol of hydrogen gas. Assume the molar volume is 22.4 litres mol$^{-1}$.**

#### Solution: Method 1

The triangle in Figure 6.1 shows how the number of moles, volume of gas and molar volume are connected.

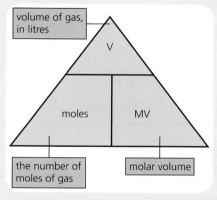

**Figure 6.1** This triangle will help with calculations about molar volume

Using the triangle, you should be able to see that:

$$V = \text{moles} \times MV \qquad \text{moles} = \frac{V}{MV} \qquad MV = \frac{V}{\text{moles}}$$

To solve this problem, we will use $V = \text{moles} \times MV$

$V = 3 \times 22.4 = $ **67.2 litres**

#### Solution: Method 2

This can also be solved using proportion.

**Step 1: Form a relationship between the number of moles and the volume, putting the volume on the right-hand side since this is the quantity you are trying to calculate.**

number of moles → volume

$$1 \text{ mol} \rightarrow 22.4 \text{ litres}$$

**Step 2: Calculate for the number of moles being asked for.**

To do this, you need to multiply both sides of the relationship by 3.

3 mol → 3 × 22.4 = **67.2 litres**

## Worked example 6.2

Calculate the number of moles of oxygen gas present in 200 $cm^3$ of the gas. Assume the molar volume is 22.4 litres $mol^{-1}$.

### Solution: Method 1

$$moles = \frac{V}{MV} = \frac{0.200}{22.4} = 0.0089 \text{ moles}$$

(Note that the volume must be changed into litres when using this equation. This is done by dividing the volume in $cm^3$ by 1000 since 1 litre is the same as 1000 $cm^3$.)

### Solution: Method 2

**volume → number of moles**

22.4 litres → 1 mol

1 litre → $\frac{1}{22.4}$ = 0.0446 moles

0.2 litres → 0.2 × 0.0446 = **0.0089 moles**

## Worked example 6.3

Calculate the molar volume for helium gas given that 4.5 moles of gas was found to occupy 108 litres at 25 °C, 1 atm.

### Solution: Method 1

$$MV = \frac{V}{moles} = \frac{108}{4.5} = 24 \text{ litres mol}^{-1}$$

### Solution: Method 2

**number of moles → volume**

4.5 mol → 108 litres

1 mol → $\frac{108}{4.5}$ = **24 litres mol$^{-1}$**

# Calculating the molar volume from experimental data

The previous examples have linked molar volume to the number of moles and volume of a gas. The examples which follow use this same link but the number of moles of gas is not given. Instead, the mass of the gas is known. To tackle these questions, you have to recognise the relationship between the number of moles and the mass of gas, i.e. 1 mol of any substance is the formula mass in grams.

## Worked example 6.4

A 200 $cm^3$ flask was filled with helium gas at 20 °C and 1 atmosphere of pressure. The flask was weighed empty and with the gas. Use the data below to calculate the molar volume of helium at this temperature and pressure.

**mass of empty flask = 84.300 g**

**mass of flask + helium = 84.334 g**

You are trying to calculate the volume of 1 mole of gas. In this question, you are not given information about the number of moles of gas but you are given information about the mass. From the data, you can calculate the mass of helium gas by subtracting the mass of the empty flask from the mass of the flask filled with helium: 84.334 – 84.300 = 0.034 g.

Since you are told that the gas was placed in a 200 $cm^3$ flask, you can assume that the volume occupied by this mass (0.0034 g) of gas is 200 $cm^3$.

### Solution: Method 1

$$\text{number of moles of helium} = \frac{mass}{gfm} = \frac{0.034}{4} = 0.0085 \text{ mol}$$

$$MV = \frac{V}{moles} = \frac{0.200}{0.0085} = 23.53 \text{ litres mol}^{-1}$$

### Solution: Method 2

**Step 1: State a relationship between mass and volume, putting volume on the right-hand side since this is the quantity you are trying to calculate**

**mass → volume**

0.034 g → 200 $cm^3$

**Step 2: Apply proportion to calculate the volume occupied by 1 g of gas. This is achieved by dividing both sides by 0.034**

1 g → $\frac{200}{0.034}$ = 5882.35 $cm^3$

**Step 3: Work out the mass of 1 mole of helium and use this mass to calculate the volume occupied by 1 mole**

1 mole of helium = 4 g

**mass → volume**

1 g → 5882.35 $cm^3$

4 g → 4 × 5882.35 = 23 529 $cm^3$ which is the same as 23.53 litres

The molar volume of helium at the temperature and pressure stated in the question is **23.53 litres mol$^{-1}$**.

## Worked example 6.5

A 100 cm³ gas syringe was filled with methane gas, $CH_4$, at 20 °C, 1 atm of pressure.

Using the data below, calculate the molar volume of methane gas at this temperature and pressure.

mass of empty gas syringe = 112.700 g

mass of gas syringe + methane gas = 112.765 g

### Solution: Method 1

number of moles of methane = $\frac{mass}{gfm}$

$= \frac{0.065}{16}$

$= 0.00406$

$MV = \frac{V}{moles}$

$= \frac{0.100}{0.00406}$

$= 24.62$ litres mol⁻¹

### Solution: Method 2

mass → volume

0.065 g → 100 cm³

1 g → $\frac{100}{0.065}$ = 1538.46 cm³

16 g → 16 × 1538.46 = 24 615 cm³

The molar volume of methane is **24.62 litres mol⁻¹**.

## Worked example 6.6

100 cm³ of a gas had a mass of 0.0179 g. Given that the molar volume of this gas was 22.4 litres mol⁻¹, calculate the mass of 1 mol of the gas. Hence, identify the gas.

### Solution: Method 1

moles = $\frac{V}{MV}$

$= \frac{0.100}{22.4}$

$= 0.0045$ mol

mass of 1 mol (gfm) = $\frac{mass}{mol}$

$= \frac{0.00179}{0.0045}$

$= 4$ g

**The gas is helium.**

### Solution: Method 2

volume → mass

100 cm³ → 0.0179 g

1 cm³ → 0.000179 g

22 400 cm³ → 22 400 × 0.000179 g = **4 g**

**The gas is helium.**

## Worked example 6.7

200 cm³ of a hydrocarbon gas had a mass of 0.1432 g. Given that the molar volume of this gas was 22.4 litres mol⁻¹, calculate the mass of 1 mol of the gas. Hence, write the molecular formula for the gas.

### Solution: Method 1

moles = $\frac{V}{MV}$

$= \frac{0.200}{22.4}$

$= 0.0089$ mol

mass of 1 mol (gfm) = $\frac{mass}{mol}$

$= \frac{0.1432}{0.0089}$

$= 16$ g

The formula must be $CH_4$.

### Solution: Method 2

volume → mass

200 cm³ → 0.1432 g

1 cm³ → 0.000716 g

22 400 cm³ → 22 400 × 0.000716 g = **16 g**

**The formula must be $CH_4$.**

## Questions

**Calculating volume, number of moles or molar volume (these question types are similar to worked examples 6.1–6.3).**

**In questions 1–8, assume the molar volume is 22.4 litres mol$^{-1}$.**

1  Calculate the volume occupied by 12 moles of $CO_2(g)$.

2  Calculate the volume occupied by 0.5 moles of $H_2(g)$.

3  Calculate the volume occupied by 0.2 moles of $Ar(g)$.

4  Calculate the volume occupied by 0.1 moles of $Ne(g)$.

5  Calculate the number of moles of $O_2$ gas present in 5 litres of the gas.

6  Calculate the number of moles of $SO_2$ gas present in 0.8 litres of the gas.

7  Calculate the number of moles of He present in 100 cm$^3$ of the gas.

8  Calculate the number of moles of CO present in 500 cm$^3$ of the gas.

9  Calculate the molar volume given that 0.04 moles of a gas was found to occupy 1 litre.

10  Calculate the molar volume given that 0.002 moles of a gas was found to occupy 45 cm$^3$.

**Calculating volume, number of moles or molar volume where the mass of a gas is used (these question types are similar to worked examples 6.4 and 6.5).**

11  At 0 °C and 1 atmosphere of pressure, a 60 cm$^3$ sample of carbon monoxide, CO, weighs 0.075 g. Calculate its molar volume under these conditions.

12  Ethyne, $C_2H_2$, has a molar volume of 22 l under certain conditions. Calculate the mass of 20 cm$^3$ of ethyne under these conditions.

13  Helium, He, has a molar volume of 54 l under certain conditions. What volume would 0.24 g of helium occupy under the same conditions?

14  A 40 cm$^3$ sample of hydrogen, $H_2$, weighs 0.0008 g under certain conditions. Calculate the molar volume of hydrogen under these conditions.

15  Under certain conditions, the molar volume of ammonia, $NH_3$, is 13.8 l. Calculate the mass of 150 l of this gas under these conditions.

16  Sulfur trioxide, $SO_3$, has a molar volume of 24 l under certain conditions. Calculate the volume which 0.15 g of the gas would occupy under these conditions.

17  A 10 cm$^3$ sample of xenon, Xe, weighs 0.054 g at room temperature and pressure. Calculate its molar volume under these conditions.

18  Under certain conditions, sulfur dioxide has a molar volume of 58.2 l. Calculate the mass of 1000 l of the gas under these conditions.

19  Butane, $C_4H_{10}$, has a molar volume of 32 l under certain conditions of temperature and pressure. Calculate the volume of 0.045 g of butane under the same conditions.

20  A 100 cm$^3$ flask is filled with carbon dioxide, $CO_2$, at room temperature and pressure. The mass of the gas in the flask is found to be 0.181 g. Calculate the molar volume of the gas under these conditions.

21  A 200 cm$^3$ gas syringe was filled with nitrogen gas, $N_2$, at 20 °C, 1 atm of pressure. Using the data below, calculate the molar volume of nitrogen gas at this temperature and pressure.

mass of empty gas syringe = 95.400 g

mass of gas syringe + nitrogen gas = 95.643 g

→

**22** A $100 \, cm^3$ flask was filled with carbon dioxide gas, $CO_2$, at 20 °C and 1 atmosphere of pressure. The flask was weighed empty and with the gas. Use the data below to calculate the molar volume of carbon dioxide at this temperature and pressure.

mass of empty flask = 56.500 g

mass of flask + helium = 56.696 g

**23** A helium balloon was found to contain $75 \, cm^3$ of helium gas at 20 °C and 1 atmosphere of pressure. Use the data below to calculate the molar volume of helium at this temperature and pressure.

mass of empty balloon = 4.560 g

mass of balloon + helium = 4.573 g

**24** A gas cylinder was found to contain 120 litres of oxygen gas, $O_2$, at 20 °C and 1 atmosphere of pressure. Use the data below to calculate the molar volume of oxygen gas at this temperature and pressure.

mass of empty cylinder = 18.000 kg

mass of cylinder + oxygen = 18.160 kg

**Using molar volume to calculate the formula mass and, hence, the identity of a gas (these question types are similar to worked examples 6.6 and 6.7).**

**25** 5.07 litres of a halogen gas was found to weigh 15 g. Assuming the molar volume was 24 litres $mol^{-1}$, calculate the formula mass of the gas and, hence, identify the gas.

**26** 0.1 g of a hydrocarbon gas was found to occupy $80 \, cm^3$. Assuming the molar volume was 24 litres $mol^{-1}$, calculate the formula mass of the gas and, hence, state the molecular formula for the gas.

**27** A sample of gas, known to be an oxide of carbon, weighing 0.5 g was found to occupy 0.27 litres. Assuming the molar volume was 24 litres $mol^{-1}$, calculate the formula mass of the gas and, hence, identify the gas.

**28** $50 \, cm^3$ of ethane, $C_2H_6$, weighs 0.0625 g.

   a) Calculate the molar volume of the gas under these conditions.

   b) Under the same conditions, 50 cm3 of an unknown gas is found to weigh 0.0792 g. Calculate the formula mass (the mass of 1 mole) of the unknown gas.

**29** Under certain conditions, a $200 \, cm^3$ sample of oxygen, $O_2$, weighs 0.358 g. A $400 \, cm^3$ sample of a gas, X, weighs 0.354 g under the same conditions. Calculate the formula mass of X.

**30** A $200 \, cm^3$ sample of hydrogen sulfide, $H_2S$, weighs 0.268 g. Under the same conditions of temperature and pressure, $40 \, cm^3$ of an unknown gas weighed 0.05 g. Calculate the formula mass of the unknown gas.

# 7 Calculations from equations using gas volumes

This chapter will study calculations from equations where the only chemicals we have to consider are gases and the quantities are expressed as volumes.

## Balanced chemical equations and volumes of gases

Consider the following balanced equation for the combustion of 1 mol of ethene ($C_2H_4$) gas:

$$C_2H_4(g) + 3O_2(g) \rightarrow 2CO_2(g) + 2H_2O(g)$$

If we had $100\,cm^3$ of ethene gas, what volume of oxygen would we require to react completely with the ethene?

The balanced equation tells us that 1 mol of ethene reacts with 3 mol of oxygen.

Since the molar volume is the same for all gases under the same conditions, we could easily calculate the volume occupied by each gas. For example, if the molar volume was 22 litres $mol^{-1}$, we could say that 22 litres of ethene would react with 66 litres of oxygen. The volume of oxygen required is three times the volume of ethene.

In other words, **the volume of gas is proportional to the number of moles**. For example, 10 mol of a gas will contain 10 times the volume of gas compared to 1 mol of the gas.

So, if we have $100\,cm^3$ of ethene, $300\,cm^3$ of oxygen is required to react.

We can use this knowledge to calculate the volume of products formed:

$100\,cm^3$ of ethene would produce $200\,cm^3$ of carbon dioxide gas and $200\,cm^3$ of water vapour.

## Worked example 7.1

Calculate the volume of carbon dioxide produced when $800\,cm^3$ of oxygen is reacted completely with an excess of carbon:

$$C(s) + O_2(g) \rightarrow CO_2(g)$$

### Solution

1 mol of oxygen will react to produce 1 mol of carbon dioxide.

$800\,cm^3$ of oxygen will react to produce **$800\,cm^3$ of carbon dioxide**.

## Worked example 7.2

Calculate the volume of carbon monoxide produced when 5 litres of oxygen are reacted with excess carbon:

$$2C(s) + O_2(g) \rightarrow 2CO(g)$$

### Solution

1 mol of oxygen will react to produce 2 mol of carbon monoxide.

5 litres of oxygen will react to produce **10 litres of carbon monoxide**.

## Worked example 7.3

Calculate the volumes of gases remaining if 50 litres of methane gas are burned in 200 litres of oxygen gas:

$$CH_4(g) + 2O_2(g) \rightarrow CO_2(g) + 2H_2O(g)$$

In this example, we are given information about two gases. We therefore have to calculate which gas is in excess and use the limiting reactant (the one not in excess) to calculate the volume of the remaining gases.

### Solution

50 litres of methane will react with 100 litres of oxygen i.e. oxygen is in excess since we have more than 100 litres (the question states that there are 200 litres of oxygen).

We can use the methane volume to calculate the volume of the products:

50 litres of methane will produce 50 litres of $CO_2(g)$ and 100 litres of $H_2O(g)$.

The volume of oxygen remaining will be (200 – 100) = 100 litres.

In summary, the volume of remaining gases will be:

$CH_4(g)$ = 0 litres

$O_2(g)$ = 100 litres

$CO_2(g)$ = 50 litres

$H_2O(g)$ = 100 litres

## Worked example 7.4

Calculate the volume and composition of remaining gases if 50 cm³ of ethene gas is reacted with 99 cm³ of oxygen gas.

$$C_2H_4(g) + 3O_2(g) \rightarrow 2CO_2(g) + 2H_2O(g)$$

### Solution

50 cm³ of ethene gas would react with 150 cm³ of oxygen gas.

Since we only have 99 cm³ of oxygen gas, the ethene is in excess. We therefore use the volume of oxygen to calculate the volume and composition of remaining gases:

99 cm³ of oxygen will react with 33 cm³ of ethene to produce 66 cm³ of carbon dioxide and 66 cm³ of water.

The composition and volume of remaining gases will be:

$C_2H_4(g)$ = 50 – 33 = 17 cm³

$O_2(g)$ = 0 cm³

$CO_2(g)$ = 66 cm³

$H_2O(g)$ = 66 cm³

## Worked example 7.5

Calculate the volume and composition of remaining gases if 80 cm³ of ethene gas was reacted with 300 cm³ of oxygen gas and the products were cooled to room temperature.

$$C_2H_4(g) + 3O_2(g) \rightarrow 2CO_2(g) + 2H_2O(l)$$

### Solution

80 cm³ of ethene would react with 240 cm³ of oxygen.

Since we have 300 cm³ of oxygen gas but only require 240 cm³, the oxygen is in excess. We therefore use the volume of ethene to calculate the volume and composition of remaining gases:

80 cm³ of ethene gas will produce 160 cm³ of carbon dioxide.

Composition and volume of remaining gases:

$C_2H_4(g)$ = 0 cm³

$O_2(g)$ = 60 cm³ (from 300 – 240)

$CO_2(g)$ = 160 cm³

Note: the question states that the products were cooled to room temperature so water is a liquid (shown as $H_2O(l)$ in the equation). We cannot calculate a volume for water since this gas volume rule only applies to gases.

## Questions

**1** $100 \, cm^3$ of methane is exploded with $300 \, cm^3$ of oxygen.

$$CH_4(g) + 2O_2(g) \rightarrow CO_2(g) + 2H_2O(l)$$

   **a)** Which gas is in excess?

   **b)** What is the volume and composition of the resulting gas mixture?

**2** $50 \, cm^3$ of carbon monoxide is burned with $20 \, cm^3$ of oxygen.

$$CO(g) + \tfrac{1}{2}O_2(g) \rightarrow CO_2(g)$$

   **a)** Which gas is in excess?

   **b)** What is the volume and composition of the resulting gas mixture?

**3** $20 \, l$ of propane is burned in $140 \, l$ of oxygen. Calculate the volume and composition of the resulting gas mixture.

$$C_3H_8(g) + 5O_2(g) \rightarrow 3CO_2(g) + 4H_2O(l)$$

**4** $400 \, l$ of cyclopropane is exploded with $5000 \, l$ of oxygen. Calculate the volume and composition of the resulting gas mixture.

$$C_3H_6(g) + 4\tfrac{1}{2}O_2(g) \rightarrow 3CO_2(g) + 3H_2O(l)$$

**5** $200 \, cm^3$ of ethane is burned completely in $2 \, l$ of oxygen. Calculate the volume and composition of the resulting gas mixture if all volume measurements were made at a temperature of $200 \, °C$ and a pressure of $1 \, atmosphere$; i.e. under conditions where water is in the gas state.

$$C_2H_6(g) + 3\tfrac{1}{2}O_2(g) \rightarrow 2CO_2(g) + 3H_2O(g)$$

**6** $100 \, l$ of hydrazine, $N_2H_4$, is burned in $400 \, l$ of oxygen to form nitrogen gas and water. What will the volume and composition of the resulting gas mixture be if all measurements were taken at $300 \, °C$ and at the same pressure?

$$N_2H_4(g) + O_2(g) \rightarrow N_2(g) + 2H_2O(g)$$

**7** $150 \, l$ of butane is burned in $2000 \, l$ of oxygen. What is the volume and composition of the resulting gas mixture? (All volume measurements are taken at a temperature of $20 \, °C$ and $1 \, atmosphere$ pressure.)

$$C_4H_{10}(g) + 6\tfrac{1}{2}O_2(g) \rightarrow 4CO_2(g) + 5H_2O(l)$$

**8** Hydrogen can be obtained from hexane by catalytic reaction with steam. Under certain conditions, $4 \times 10^4 \, l$ of hexane is reacted with an excess of steam. Assuming the reaction goes to completion, calculate the volume produced of

   **a)** carbon dioxide

   **b)** hydrogen.

$$C_6H_{14}(g) + 6H_2O(g) \rightarrow 6CO_2(g) + 13H_2(g)$$

**9** Tetrachloromethane, $CCl_4$, can be made industrially by the following process in which carbon disulfide is reacted with chlorine gas. Under certain conditions, $5 \times 10^5 \, l$ of carbon disulfide gas is reacted completely in the presence of $2.5 \times 10^6 \, l$ of chlorine. What is the composition by volume of the resulting gas mixture?

$$CS_2(g) + 3Cl_2(g) \rightarrow CCl_4(g) + S_2Cl_2(g)$$

**10** Hexane can be partially oxidised to carbon monoxide and oxygen gases. Under certain conditions, $3.15 \times 10^6 \, l$ of hydrogen gas were obtained from this reaction. What volume of gaseous hexane, measured under the same conditions, must have reacted?

$$C_6H_{14}(g) + 3O_2(g) \rightarrow 6CO(g) + 7H_2(g)$$

# 8 Calculations from equations using molar volume

In Chapters 4 and 5, we considered calculations from equations where the quantities of reactants or products were expressed as masses, or concentrations and volume for solutions. In this chapter, we consider calculations where gases are reacted or produced. The quantities will be expressed in terms of the volume of gases, requiring a use of molar volume to solve the problems. This is illustrated in the following worked examples.

## Worked example 8.1

8.8 g of propane gas, $C_3H_8$, was burned in an excess of oxygen gas. Calculate the volume of carbon dioxide produced, assuming that the molar volume is 22.4 litres mol⁻¹.

$$C_3H_8(g) + 5O_2(g) \rightarrow 3CO_2(g) + 4H_2O(g)$$

**Solution**

**Step 1: Calculate the number of moles of the reagent(s)**

In this case, the only reagent we can consider is propane since this is the only reagent we have a mass for.

$$\text{moles} = \frac{\text{mass}}{\text{gfm}}$$

$$= \frac{8.8}{44}$$

$$= 0.2 \, \text{mol}$$

**Step 2: Use the mole ratio to calculate the number of moles of product**

From the equation, 1 mol of propane produces 3 mol of carbon dioxide.

So, 0.2 mol $C_3H_8 \rightarrow$ 0.6 mol $CO_2$

**Step 3: Convert moles to volume**

$$\text{volume} = \text{mol} \times \text{MV}$$

$$= 0.6 \times 22.4$$

$$= 13.44 \, \text{litres}$$

## Worked example 8.2

4.4 g of propane gas, $C_3H_8$, was burned in 12 litres of oxygen gas. Calculate the volume of carbon dioxide produced, assuming that the molar volume is 22.4 litres mol⁻¹.

$$C_3H_8(g) + 5O_2(g) \rightarrow 3CO_2(g) + 4H_2O(g)$$

**Solution**

This question can be solved using the same approach as the previous worked example. The only difference is that you have to calculate the number of moles of both reactants to allow you to calculate which reactant is in excess.

**Step 1: Calculate the number of moles of the reagent(s)**

$$\text{number of moles of propane gas} = \frac{\text{mass}}{\text{gfm}} = \frac{4.4}{44} = 0.1 \, \text{mol}$$

$$\text{number of moles of oxygen gas} = \frac{\text{volume}}{\text{molar volume}}$$

$$= \frac{12}{22.4}$$

$$= 0.54 \, \text{mol}$$

According to the equation, 1 mol of propane will react with 5 mol of oxygen.

Therefore, 0.1 mol of propane will react with 0.5 mol of oxygen.

Since we have 0.54 mol of oxygen, the oxygen is in excess.

**Step 2: Use the mole ratio to calculate the number of moles of product**

From the equation, 1 mol of propane produces 3 mol of carbon dioxide.

So, 0.1 mol $C_3H_8 \rightarrow$ 0.3 mol $CO_2$

**Step 3: Convert moles to volume**

$$\text{volume} = \text{mol} \times \text{MV}$$

$$= 0.3 \times 22.4$$

$$= 6.72 \, \text{litres}$$

# Chapter 8

## Worked example 8.3

Calcium carbonate decomposes according to the following equation:

$$CaCO_3(s) \rightarrow CaO(s) + CO_2(g)$$

In an experiment, 140 cm³ of carbon dioxide gas was collected from the decomposition of a sample of calcium carbonate. Calculate the mass of calcium carbonate that decomposed. Assume molar volume is 22.4 litres mol⁻¹.

### Solution

**Step 1: Calculate the number of moles of the product**

In this case, the only chemical we can consider is carbon dioxide since this is the only chemical we have a quantity for.

$$\text{moles} = \frac{\text{volume}}{\text{molar volume}}$$

$$= \frac{140}{22\,400}$$

$$= 6.25 \times 10^{-3}\,\text{mol}$$

**Step 2: Use the mole ratio to calculate the number of moles of reactant**

From the equation, 1 mol of carbon dioxide was produced from 1 mol of calcium carbonate.

So, $6.25 \times 10^{-3}$ mol $CO_2 \rightarrow 6.25 \times 10^{-3}$ mol $CaCO_3$

**Step 3: Convert moles to mass**

mass of $CaCO_3$ = mole × gfm

$$= 6.25 \times 10^{-3} \times 100$$

$$= 0.625\,\text{g}$$

## Questions

1. 7.2 g of methane is burned completely in oxygen according to the following equation. Calculate the volume of carbon dioxide produced. Assume molar volume is 22 litres mol⁻¹.

$$CH_4(g) + 2O_2(g) \rightarrow CO_2(g) + 2H_2O(g)$$

2. 5.3 g of sodium carbonate is reacted with an excess of dilute hydrochloric acid. Calculate the volume of carbon dioxide produced. Assume molar volume is 22 litres mol⁻¹.

$$Na_2CO_3(s) + 2HCl(aq) \rightarrow 2NaCl(aq) + CO_2(g) + H_2O(l)$$

3. 400 l of oxygen gas was used to react with ammonia gas. Calculate the volume of ammonia reacting with the oxygen gas.

$$4NH_3(g) + 5O_2(g) \rightarrow 4NO(g) + 6H_2O(l)$$

4. 1824 g of fluorine is reacted with an excess of steam. Calculate the volume of ozone gas, $O_3$, produced. Assume molar volume is 22 litres mol⁻¹.

$$3F_2(g) + 3H_2O(g) \rightarrow 6HF(g) + O_3(g)$$

5. Calculate the mass of zinc required to react with an excess of hydrochloric acid to produce 4 l of hydrogen gas. Assume molar volume is 22 litres mol⁻¹.

$$Zn(s) + 2HCl(aq) \rightarrow ZnCl_2(aq) + H_2(g)$$

6. Calculate the volume of carbon dioxide produced when 20 g of calcium carbonate is reacted with 100 cm³ of 1 mol l⁻¹ sulfuric acid. Assume molar volume is 22 litres mol⁻¹.

$$H_2SO_4(aq) + CaCO_3(s) \rightarrow CaSO_4(aq) + CO_2(g) + H_2O(l)$$

7. 12.8 g of hydrazine, $N_2H_4$, was reacted with 10 litres of fluorine gas. Calculate the volume of hydrogen fluoride gas produced. Assume molar volume is 22 litres mol⁻¹.

$$N_2H_4(g) + 2F_2(g) \rightarrow N_2(g) + 4HF(g)$$

8. 1500 l carbon dioxide was obtained through the reduction of iron(III) oxide by carbon monoxide. What mass of iron(III) oxide must have been reduced? Assume molar volume is 22 litres mol⁻¹.

$$Fe_2O_3(s) + 3CO(g) \rightarrow 2Fe(s) + 3CO_2(g)$$

9. 15.24 g of copper is reacted with 200 cm³ of 4 mol l⁻¹ nitric acid. Calculate the volume of NO(g) produced. Assume molar volume is 22 litres mol⁻¹.

$$3Cu(s) + 8HNO_3(aq) \rightarrow 3Cu(NO_3)_2(aq) + 4H_2O(l) + 2NO(g)$$

10. $1.08 \times 10^6$ l of hexane, $C_6H_{14}$, was reacted with an excess of steam according to the following equation. If the molar volume of the hexane under these conditions is 120 l, calculate the mass of hydrogen which would be produced.

$$C_6H_{14}(g) + 6H_2O(g) \rightarrow 6CO(g) + 13H_2(g)$$

# 9 Percentage yield and using percentages in calculations

In many chemical reactions, not all reactants are converted into the desired products. There are a number of reasons for this, with the most common reasons listed below. All of these will result in a lower than expected yield of product(s).

- Many reactions are reversible. This means that products can break down into reactants.

- There may be side reactions: sometimes reactants can form products that are not desired or expected. In addition, products can react with reactants to form completely different products.

- The reactants may be impure.

- The reactants may break down (decompose) under the conditions used in the reaction.

## Percentage yield

The **percentage yield** is a simple way of comparing the amount of product *actually* obtained from a reaction with the amount *expected*. This is illustrated by the following example.

Consider the reaction between hydrogen and chlorine to produce hydrogen chloride.

$$H_2 + Cl_2 \rightarrow 2HCl$$

If 20 g of hydrogen was to react with an excess of chlorine, we could calculate the mass of hydrogen chloride formed using the methods detailed in previous chapters:

number of moles of hydrogen $= \frac{20}{2} = 10$ moles

From the equation, 1 mol of $H_2$ reacts to produce 2 mol of HCl.

So, 10 mol of $H_2$ would react to produce 20 mol of HCl.

mass of HCl expected $= 20 \times 36.5 = \textbf{730 g}$

In summary, if we get complete conversion of the 20 g of hydrogen into hydrogen chloride, we would expect to obtain 730 g of hydrogen chloride. This is known as the theoretical yield, i.e. what you would expect to be formed assuming 100% conversion of reactants into products.

If, however, you were told that the reaction had a 50% yield, this tells us that only 50% of the expected product is formed. In other words, 50% of 730 = **365 g** would be formed.

The percentage yield can be used to calculate the actual quantity of product formed, as outlined in the above example.

A simple expression which allows you to calculate the percentage yield is:

$$\textbf{percentage yield} = \frac{\textbf{actual yield}}{\textbf{theoretical yield}} \times \textbf{100}$$

- The **actual yield** is the quantity you obtain from experiment.

- The **theoretical yield** is what you would expect to obtain by calculation, assuming 100% conversion of reactants into products.

### Worked example 9.1

**20 g of hydrogen gas was reacted with excess chlorine and produced 600 g of hydrogen chloride.**

$$\textbf{H}_2 + \textbf{Cl}_2 \rightarrow \textbf{2HCl}$$

**Calculate the percentage yield.**

**Solution**

percentage yield $= \frac{\text{actual yield}}{\text{theoretical yield}} \times 100$

From the previous calculation, the theoretical yield of HCl from 20 g $H_2$ would be 730 g.

The actual yield (from the statement in the question) is 600 g.

percentage yield $= \frac{600}{730} \times 100$

$= \textbf{82.2\%}$

## Worked example 9.2

160 g of methane gas was reacted with an excess of oxygen. 320 g of carbon dioxide gas was produced. Calculate the percentage yield.

$$CH_4(g) + 2O_2(g) \rightarrow CO_2(g) + 2H_2O(g)$$

### Solution

moles of methane $= \dfrac{mass}{gfm}$

$\qquad = \dfrac{160}{16}$

$\qquad = 10 \, mol$

10 mol of methane $\rightarrow$ 10 mol of carbon dioxide

mass of carbon dioxide = mole × gfm

$\qquad = 10 \times 44$

$\qquad = 440 \, g$

(this is the theoretical yield)

percentage yield $= \dfrac{actual \, yield}{theoretical \, yield} \times 100$

$\qquad = \dfrac{320}{440} \times 100$

$\qquad = \mathbf{72.7\%}$

## Worked example 9.3

20 g of methane gas was reacted with an excess of oxygen. Assuming an 80% percentage yield, calculate the mass of carbon dioxide gas produced.

$$CH_4(g) + 2O_2(g) \rightarrow CO_2(g) + 2H_2O(g)$$

### Solution

moles of methane $= \dfrac{mass}{gfm}$

$\qquad = \dfrac{20}{16}$

$\qquad = 1.25 \, mol$

1.25 mol of methane $\rightarrow$ 1.25 mol of carbon dioxide

mass of carbon dioxide = mole × gfm

$\qquad = 1.25 \times 44$

$\qquad = 55 \, g$

(this is the theoretical yield)

actual yield = % yield × theoretical yield

$\qquad = 80\% \times 55$

$\qquad = \mathbf{44 \, g}$

## Questions

1  Carbon monoxide and hydrogen gas can be reacted to form methanol, $CH_3OH$. 22.4 g of carbon monoxide was reacted with an excess of hydrogen to produce 19.2 g of methanol. Calculate the percentage yield.

$$CO + 2H_2 \rightarrow CH_3OH$$

2  Ethene can react with hydrogen iodide to form iodoethane, $C_2H_5I$. 7 g of ethene reacted to produce 15.59 g of iodoethane. Calculate the percentage yield.

$$C_2H_4 + HI \rightarrow C_2H_5I$$

3  5.072 g of iron(II) chloride reacted to form 4.869 g of iron(III) chloride. Calculate the percentage yield.

$$2FeCl_2 + Cl_2 \rightarrow 2FeCl_3$$

4  18.72 g of benzene, $C_6H_6$, reacted with excess nitric acid to form 22.14 g of nitrobenzene, $C_6H_5NO_2$. Calculate the percentage yield.

$$C_6H_6 + HNO_3 \rightarrow C_6H_5NO_2 + H_2O$$

5  5.3 g of sodium carbonate was reacted with an excess of sulfuric acid, resulting in 1.76 g of carbon dioxide being given off. Calculate the percentage yield.

$$Na_2CO_3 + H_2SO_4 \rightarrow Na_2SO_4 + CO_2 + H_2O$$

6  $3.182 \times 10^3$ kg of copper(I) sulfide reacted with oxygen to produce $1.716 \times 10^3$ kg of copper(I) oxide. Calculate the percentage yield.

$$2Cu_2S + 3O_2 \rightarrow 2Cu_2O + 2SO_2$$

7  $1.36 \times 10^3$ kg of ammonia was reacted and produced $1.68 \times 10^3$ kg of nitrogen monoxide. Calculate the percentage yield.

$$4NH_3 + 5O_2 \rightarrow 4NO + 6H_2O$$

8  $3.205 \times 10^3$ kg of sulfur dioxide reacted to form $2.403 \times 10^3$ kg of sulfur trioxide. Calculate the percentage yield.

$$2SO_2 + O_2 \rightarrow 2SO_3$$

# Using percentages in calculations

Another question type worth considering in the Higher course is where you have to calculate how much product is formed when not all of the reactant is used up. The method for this is exactly the same as tackling any calculation from an equation, except that you have to calculate the quantity reacting before you can start the calculation. This is illustrated in the following worked examples.

---

**Worked example 9.4**

17.5 g of ethene gas was reacted with an excess of chlorine gas to produce dichloroethene:

$$C_2H_4 + Cl_2 \rightarrow C_2H_4Cl_2$$

If only 80% of the ethene reacts, calculate the mass of dichloroethene produced.

**Solution**

You have to base your calculation on the actual mass of ethene reacting: 80% of 17.5 g = 14 g

$$\text{moles of ethene} = \frac{\text{mass}}{\text{gfm}}$$

$$= \frac{14}{28}$$

$$= 0.5\ \text{mol}$$

1 mol of ethene $\rightarrow$ 1 mol of dichloroethene

0.5 mol of ethene $\rightarrow$ 0.5 mol of dichloroethene

mass of dichloroethene = mole × gfm

$$= 0.5 \times 99$$

$$= \textbf{49.5 g}$$

---

**Worked example 9.5**

A 10 g sample of rock containing calcium carbonate was reacted with excess hydrochloric acid:

$$CaCO_3 + 2HCl \rightarrow CaCl_2 + H_2O + CO_2$$

Assuming that the rock contains 85% calcium carbonate, calculate the volume of carbon dioxide released.

(Take molar volume to be 22.4 litres mol$^{-1}$.)

**Solution**

mass of calcium carbonate reacting = 85% of 10 g

$$= 8.5\ \text{g}$$

$$\text{moles of calcium carbonate} = \frac{8.5}{100}$$

$$= 0.085\ \text{mol}$$

1 mol of calcium carbonate $\rightarrow$ 1 mol of carbon dioxide

0.085 mol of calcium carbonate $\rightarrow$ 0.085 mol of carbon dioxide

volume = moles × molar volume

$$= 0.085 \times 22.4$$

$$= \textbf{1.904 litres}$$

## Questions

9   19.95 g of iron(III)oxide was reacted with carbon monoxide. Assuming only 80% of the iron(III) oxide reacts, calculate the mass of iron produced.

$Fe_2O_3 + 3CO \rightarrow 2Fe + 3CO_2$

10  A rock containing 60% calcium carbonate by mass was crushed and heated until all the calcium carbonate had decomposed according to the following equation. Calculate the mass of calcium oxide formed from 40 g of the rock.

$CaCO_3 \rightarrow CaO + CO_2$

11  4 g of methane was reacted to produce hydrogen gas. Assuming only 30% of the methane reacted, calculate the mass of hydrogen formed.

$CH_4 + H_2O \rightarrow 3H_2 + CO$

12  47.86 g of lead sulfide was reacted with excess oxygen. Assuming a 75% yield, calculate the mass of lead(II) oxide formed.

$2PbS + 3O_2 \rightarrow 2PbO + 2SO_2$

13  37.98 g of titanium(IV) chloride was reacted with excess magnesium. Assuming a 60% yield, calculate the mass of titanium formed.

$TiCl_4 + 2Mg \rightarrow Ti + 2MgCl_2$

14  In the manufacture of ammonia by the Haber Process, under certain conditions there is only a 70% yield of product. If $8.40 \times 10^4$ kg of nitrogen is reacted with an excess of hydrogen under these conditions, calculate the mass of ammonia produced.

$N_2 + 3H_2 \rightarrow 2NH_3$

15  The fermentation of glucose to ethanol and carbon dioxide produces a 75% yield of ethanol under certain conditions. If 72 kg of glucose is fermented under these conditions, what mass of ethanol would be produced?

$C_6H_{12}O_6 \rightarrow 2C_2H_5OH + 2CO_2$

16  $7.68 \times 10^4$ kg of sulfur dioxide was reacted with oxygen to produce sulfur trioxide. Assuming the reaction goes to 75% completion, calculate the mass of sulfur trioxide formed.

$2SO_2 + O_2 \rightarrow 2SO_3$

17  $1.95 \times 10^4$ kg of a zinc ore, which is known to contain 40% zinc sulfide by mass, is reacted with excess oxygen to produce zinc oxide. Calculate the mass of zinc oxide formed.

$2ZnS + 3O_2 \rightarrow 2ZnO + 2SO_2$

18  750 kg of ethanoic acid, $CH_3COOH$, was reacted with methanol to produce the ester methyl ethanoate. If there is a 72% yield of product, calculate the mass of methyl ethanoate formed.

$CH_3COOH + CH_3OH \rightarrow CH_3COOCH_3 + H_2O$

19  A 4.65 kg batch of scrap metal, consisting mainly of iron, is analysed by being treated with an excess of hydrochloric acid, causing all the iron to be converted to iron(II) chloride solution. After evaporation of the solution and the removal of all other substances, 9.51 kg of pure, solid, iron(II) chloride was obtained.

$Fe + 2HCl \rightarrow FeCl_2 + H_2$

a)  Calculate the mass of pure iron in the batch of scrap.

b)  Express the mass of pure iron in the scrap metal as a percentage of the mass of the original batch.

20  $6.65 \times 10^4$ kg of an iron ore which is impure iron(III) oxide is reacted with an excess of carbon monoxide, producing $2.79 \times 10^4$ kg of iron.

$Fe_2O_3 + 3CO \rightarrow 2Fe + 3CO_2$

a)  Calculate the mass of pure iron(III) oxide in the ore.

b)  Calculate the percentage, by mass, of iron(III) oxide in the ore.

# 10 Atom economy

Chemists work with chemical reactions to make new products which are useful and, in most cases, can generate maximum profit. One way of assessing a chemical reaction is to use the percentage yield. Reactions with a high percentage yield are highly desirable. However, reactions can have a high percentage yield but can still be wasteful if they produce lots of other products. For example, when calcium carbonate reacts with hydrochloric acid, three products are formed:

$$CaCO_3 + 2HCl \rightarrow CaCl_2 + H_2O + CO_2$$

If the desired product is calcium chloride, even if the reaction has a 100% yield, the other products (water and carbon dioxide) must be considered as they may affect the purity of the final product or may affect the actual process used to make the product.

Another method of assessing a chemical reaction is to use the concept of **atom economy**. This allows us to calculate how much of the reactant is found in the product.

## Worked example 10.1

Calculate the atom economy for the production of calcium chloride from the following reaction:

$$CaCO_3 + 2HCl \rightarrow CaCl_2 + H_2O + CO_2$$

**Solution**

The desired product is $CaCl_2$ which has a gfm of $40 + (2 \times 35.5) = 111\,g$

The total mass of reactants = $CaCO_3$ (40 + 12 + 48) + $2HCl$ (2 × 36.5) = 173 g

atom economy = $\dfrac{\text{mass of desired product(s)}}{\text{total mass of reactants}} \times 100$

$= \dfrac{111}{173} \times 100 = 64\%$

This tells us that 64% of the reactant atoms appear in the desired product. The other 36% of reactant atoms are used to make the water and carbon dioxide.

Note: when calculating the mass of desired product(s) or reactant(s), you must take into account the number of moles of each species. In this case, we had to multiply the mass of HCl by 2 since two HCl molecules are required for complete reaction.

## Calculating the atom economy

The atom economy can be calculated using the expression:

**atom economy** = $\dfrac{\text{mass of desired product(s)}}{\text{total mass of reactants}} \times \textbf{100}$

The 'mass' is taken as the gfm. This is shown in the following worked examples.

## Worked example 10.2

Calculate the atom economy for the production of carbon dioxide from the following reaction:

$$C_2H_4 + 3O_2 \rightarrow 2CO_2 + 2H_2O$$

**Solution**

The desired product is $CO_2$. Total mass = 2 × 44 = 88 g

The total mass of reactants = $C_2H_4$ (28) + $3O_2$ (3 × 32)

$= 124\,g$

atom economy = $\dfrac{\text{mass of desired product(s)}}{\text{total mass of reactants}} \times 100$

$= \dfrac{88}{124} \times 100$

$= 71\%$

## Worked example 10.3

Calculate the atom economy for the production of hydrogen chloride from the following reaction:

$$H_2 + Cl_2 \rightarrow 2HCl$$

**Solution**

The desired product is HCl. Total mass = 2 × 36.5 = 73 g

The total mass of reactants = $H_2$ (2) + $Cl_2$ (71) = 73 g

atom economy = $\dfrac{\text{mass of desired product (s)}}{\text{total mass of reactants}} \times 100$

$= \dfrac{73}{73} \times 100$

$= 100\%$

In the final worked example, we encounter a reaction which is highly desirable, i.e. one with 100% atom economy. This means that all reactant atoms are converted to the desired product. Any reaction which has only one product will always have 100% atom economy.

# Chapter 10

## Questions

1 Calculate the atom economy for the production of iron.

$Fe_2O_3 + 3CO \rightarrow 2Fe + 3CO_2$

2 Calculate the atom economy for the production of calcium oxide.

$CaCO_3 \rightarrow CaO + CO_2$

3 Calculate the atom economy for the production of hydrogen.

$CH_4 + H_2O \rightarrow 3H_2 + CO$

4 Calculate the atom economy for the production of lead(II) oxide.

$2PbS + 3O_2 \rightarrow 2PbO + 2SO_2$

5 Calculate the atom economy for the production of titanium.

$TiCl_4 + 2Mg \rightarrow Ti + 2MgCl_2$

6 Calculate the atom economy for the production of ammonia.

$N_2 + 3H_2 \rightarrow 2NH_3$

7 Calculate the atom economy for the production of ethanol.

$C_6H_{12}O_6 \rightarrow 2C_2H_5OH + 2CO_2$

8 Calculate the atom economy for the production of sulfur trioxide.

$2SO_2 + O_2 \rightarrow 2SO_3$

9 Calculate the atom economy for the production of zinc(II) oxide.

$2ZnS + 3O_2 \rightarrow 2ZnO + 2SO_2$

10 Calculate the atom economy for the production of methylethanoate.

$CH_3COOH + CH_3OH \rightarrow CH_3COOCH_3 + H_2O$

11 Show, by calculation, that the atom economy for the production of methanol is 100%.

$CO + 2H_2 \rightarrow CH_3OH$

12 Calculate the atom economy for the production of copper(I) oxide.

$2Cu_2S + 3O_2 \rightarrow 2Cu_2O + 2SO_2$

# 11 Energy changes

In designing a chemical process to make new products, chemists need to know about the energy released or taken in during a chemical reaction. This chapter will examine how to calculate this energy change and will apply this to the calculation of the enthalpy of combustion.

## The enthalpy change

The heat energy taken in or released in a chemical reaction can be obtained experimentally by measuring the change in temperature ($\Delta T$) of a certain mass of water ($m$) caused by a chemical reaction. The expression used is:

$$E_h = cm\Delta T$$

- $c$ is the specific heat capacity of water = $4.18\,kJ\,kg^{-1}\,°C^{-1}$. This value can be found in the data booklet. It refers to the fact that if $4.18\,kJ$ of heat is added to (or taken out of) $1\,kg$ of water, the temperature will rise (or fall) by $1\,°C$.

- $m$ is the mass of water (or aqueous solution) being heated (or cooled down), expressed in kg. Often, the amount of water is expressed as a volume, in $cm^3$ or litres. To change volumes of water into mass, we use the conversion that $1\,cm^3$ of water weighs approximately $1\,g$.

- $E_h$ is the enthalpy change for the reaction, measured in kJ, i.e. the heat energy released or taken in. Where this energy change is used to calculate the heat released, or taken in, per mole of substance, the units of $kJ\,mol^{-1}$ are used and the symbol $\Delta H$ is used.

- $\Delta T$ is the temperature change in °C of the water. This is expressed as a positive value, whether or not the change is an increase or decrease in temperature.

- Reactions which are exothermic have a negative enthalpy change, e.g. $\Delta H = -40\,kJ\,mol^{-1}$. This tells us that $40\,kJ\,mol^{-1}$ is 'lost' from the reaction, causing the surroundings to gain $40\,kJ$ of energy.

- Reactions which are endothermic have a positive enthalpy change, e.g. $\Delta H = 20\,kJ\,mol^{-1}$. This tells us that $20\,kJ\,mol^{-1}$ is gained from the surroundings, causing the surroundings to 'lose' $20\,kJ$ of energy.

### Worked example 11.1

**Calculate the quantity of energy, in kJ, required to raise the temperature of $300\,cm^3$ of water by $5\,°C$.**

**Solution**

$m = 0.3\,kg$

$c = 4.18\,kJ\,kg^{-1}\,°C^{-1}$

$\Delta T = 5\,°C$

$E_h = ?$

We fit these figures into the equation:
$E_h = cm\Delta T = 4.18 \times 0.3 \times 5 = \mathbf{6.27\,kJ}$

### Worked example 11.2

**$10\,kJ$ of energy was used to heat $100\,cm^3$ of water, which was at $20\,°C$. Calculate the highest temperature reached by the water.**

**Solution**

$m = 0.1\,kg$

$c = 4.18\,kJ\,kg^{-1}\,°C^{-1}$

$\Delta T = ?$

$E_h = 10\,kJ$

For this example, we need to rearrange the equation ($E_h = cm\Delta T$) to find $\Delta T$, the temperature change.

i.e. $\Delta T = \dfrac{E_h}{cm} = \dfrac{10}{4.18 \times 0.1} = 23.92\,°C$

This tells us that the temperature rose by $23.92\,°C$. Since the starting temperature was $20\,°C$, the highest temperature reached would be $20 + 23.92 = \mathbf{43.92\,°C}$.

## Worked example 11.3

A volume of water at 22 °C was heated with 28 kJ of energy causing the temperature to rise to 28 °C. Calculate the volume of water heated.

### Solution

$m = ?$

$c = 4.18\,kJ\,kg^{-1}\,°C^{-1}$

$\Delta T = 6$

$E_h = 28\,kJ$

For this example, we need to rearrange the equation $(E_h = cm\Delta T)$ to find $m$, the mass of water.

i.e. $m = \dfrac{E_h}{c\Delta T} = \dfrac{28}{4.18 \times 6} = 1.12\,kg$

Using the conversion that 1 kg is approximately equal to 1 litre, the volume of water = **1.12 litres**.

## Questions

1 Calculate the quantity of energy required, in kJ, to raise the temperature of 2 litres of water by 15 °C.

2 Calculate the quantity of energy required, in kJ, to raise the temperature of 200 cm³ of water by 10 °C.

3 Calculate the quantity of energy required, in kJ, to raise the temperature of 50 cm³ of water by 20 °C.

4 Calculate the quantity of energy required, in kJ, to raise the temperature of 150 cm³ of water by 2 °C.

5 80 kJ of energy was used to heat 200 cm³ of water. Calculate the temperature change of the water.

6 100 kJ of energy was used to heat 500 cm³ of water, which was at 22 °C. Calculate the highest temperature reached by the water.

7 20 kJ of energy was used to heat 400 cm³ of water. The highest temperature reached was 48 °C. Calculate the initial temperature of the water.

8 15 kJ of energy was used to heat 200 cm³ of water. The highest temperature reached was 65 °C. Calculate the initial temperature of the water.

9 A mass of water at 22 °C was heated with 15 kJ of energy causing the temperature to rise to 37 °C. Calculate the mass of water heated.

10 A mass of water at 29 °C was heated with 11 kJ of energy causing the temperature to rise to 41 °C. Calculate the mass of water heated.

## The enthalpy of combustion

The enthalpy of combustion is the amount of heat given out when 1 mol of a substance is burned completely in an excess of oxygen.

Values for enthalpy of combustion can be obtained from the data booklet. For example, the enthalpy of combustion of methane is $-891\,kJ\,mol^{-1}$, which refers to the following reaction:

$$CH_4 + 2O_2 \rightarrow CO_2 + 2H_2O \quad \Delta H = -891\,kJ\,mol^{-1}$$

The enthalpy of combustion of a flammable liquid, such as an alcohol, can be measured using the apparatus shown in Figure 11.1.

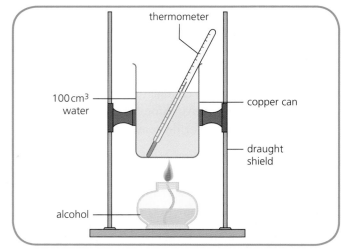

**Figure 11.1** Lab apparatus used to measure the enthalpy of combustion of a fuel

- A known mass of water is measured into a metal can and the temperature of the water is taken.

- The burner containing alcohol is weighed, lit and placed under the can.

- After some time, when the water has heated up sufficiently, the flame is put out and the burner is reweighed. The mass of alcohol burned is obtained by subtracting the final mass of the burner from its initial mass.

- The highest temperature that the water reaches is then taken. (It takes some time for all the heat in the metal can to get to the water.)

- The heat produced by burning the alcohol can be calculated using the equation $E_h = cm\Delta T$.

- The amount of heat which would have been produced had 1 mol of the alcohol been burned can then be calculated by simple proportion, knowing its formula, and therefore, the mass of 1 mol.

It should be noted that this is a very inaccurate method of measuring the enthalpy of combustion of a substance because so much heat is lost to the surroundings. Data booklet values are obtained by much more accurate methods, such as using a calorimeter, which reduce heat loss to a minimum.

## Worked example 11.4

0.02 mol of an alcohol is burned in a spirit burner. The heat given out is used to raise the temperature of 0.4 kg of water by 10 °C. Calculate the enthalpy of combustion of the alcohol.

### Solution

The enthalpy of combustion of a compound is the quantity of heat given out when 1 mol of the compound is burned completely.

First, we calculate the quantity of heat produced using $E_h = cm\Delta T$:

$m = 0.4$ kg

$c = 4.18$ kJ kg$^{-1}$ °C$^{-1}$

$\Delta T = 10$ °C

$E_h = cm\Delta T = 4.18 \times 0.4 \times 10 = 16.27$ kJ

This is the amount of heat given out when 0.02 mol of the compound is burned. We now calculate the amount of heat which would have been given out if 1 mol had been burned.

0.02 mol → 16.72 kJ of heat

$\quad$ 1 mol → $\frac{16.72}{0.02}$ = 836 kJ

So the enthalpy of combustion of the substance is **−836 kJ mol⁻¹**.

Note: the negative sign is included to show that this is heat given out; in other words, it is an exothermic reaction.

## Worked example 11.5

0.16 g of methanol, $CH_3OH$, is burned in a spirit burner. The heat from this combustion causes the temperature of 0.1 kg of water to be raised from 20 °C to 27 °C. Use this information to calculate the enthalpy of combustion of methanol.

### Solution

$E_h = cm\Delta T = 4.18 \times 0.1 \times 7 = 2.926$ kJ

The calculated value of 2.926 kJ is the quantity of heat which is given out when 0.16 g of methanol is burned.

We now calculate the heat which would have been given out if 1 mol of methanol had been burned.

The formula of methanol is $CH_3OH$. So 1 mol of methanol = 32 g.

0.16 g → 2.926 kJ

$\quad$ 1 g → $\frac{2.926}{0.16}$ = 18.29 kJ

$\quad$ 32 g → 18.29 × 32 = 585.2 kJ

So, the enthalpy of combustion of methanol = **−585.2 kJ mol⁻¹**.

# Chapter 11

## Worked example 11.6

An experiment was carried using an ethanol, $C_2H_5OH$, spirit burner to heat some water.

Use the data shown to calculate the enthalpy of combustion of ethanol.

| | |
|---|---|
| Mass of burner + ethanol before burning | 87.29 g |
| Mass of burner + ethanol after burning | 86.84 g |
| Volume of water heated | 100 cm³ |
| Temperature of water before heating | 22 °C |
| Highest temperature of water after heating | 46 °C |

**Solution**

$m = 0.1\,kg$ (from 100 cm³ of water = 100 g)

$\Delta T = 24\,°C$ (from 46 − 22)

mass of ethanol burned = 0.45 g (from 87.29 − 86.84)

$E_h = cm\Delta T = 4.18 \times 0.1 \times 24 = 10.03\,kJ$

The formula of ethanol is $C_2H_5OH$.

So 1 mol of ethanol = 46 g.

$0.45\,g \rightarrow 10.03\,kJ$

$$1\,g \rightarrow \frac{10.03}{0.45} = 22.29\,kJ$$

$$46\,g \rightarrow 22.29 \times 46 = 1025.49\,kJ$$

So, the enthalpy of combustion of ethanol
**= −1025.49 kJ mol⁻¹**

## Questions

11 A certain mass of alcohol is burned in a spirit burner which is used to heat up 0.1 kg of water in a metal can. The temperature of the water rises by 8 °C. What quantity of heat has been given out?

12 A 2 °C fall in temperature is recorded when a substance is dissolved in 0.25 kg of water. What quantity of heat has been absorbed by the water?

13 A Bunsen burner is used to heat 0.5 kg of water from 20.5 °C to 39.5 °C. How much heat has been produced in the burning of the gas?

14 0.02 mol of a hydrocarbon is burned completely in air and the heat produced is used to heat 0.1 kg of water from 20.5 °C to 29.5 °C. Calculate the enthalpy of combustion of the fuel.

15 0.025 mol of a fuel was used to heat 0.4 kg of water, which had an initial temperature of 20 °C. After heating, the temperature of the water rose to 25 °C. Calculate the enthalpy of combustion of the fuel.

16 A can containing 0.1 kg of water at 21 °C is heated to 29 °C when a burner containing an alcohol is lit underneath it. If 0.02 mol of the alcohol is burned in the process, calculate its enthalpy of combustion.

17 0.05 mol of a compound was burned and the heat released was used to heat 0.5 kg of water. The temperature of the water rose from 21 °C to 29.5 °C. Calculate the enthalpy of combustion of the compound.

18 0.32 g of methanol, $CH_3OH$, was burned in a spirit burner and used to heat up 0.2 kg of water from 19.5 °C to 27.5 °C. Calculate the enthalpy of combustion of the methanol.

19 0.2 g of methane, $CH_4$, was burned and the heat released used to raise the temperature of 0.25 kg of water from 18.5 °C to 28.5 °C. Calculate the enthalpy of combustion of methane.

20 A burner containing ethanol, $C_2H_5OH$, was used to heat up 0.4 kg of water from 21 °C to 37 °C. In the process, 0.92 g of ethanol was burned. Calculate the enthalpy of combustion of ethanol.

21 0.22 g of propane, $C_3H_8$, was burned to heat 0.25 kg of water from 21 °C to 31 °C. Calculate the enthalpy of combustion of propane.

22 A gas burner containing butane, $C_4H_{10}$, was used to heat 0.15 kg of water from 22.5 °C to 31 °C. 0.116 g of butane was burned in the process. Calculate the enthalpy of combustion of butane.

# Enthalpy of combustion: more complex examples

The next two worked examples involve using known values for the enthalpy of combustion to calculate other values. These are slightly more complex and need more arithmetical manipulation than the previous calculations in this chapter.

## Worked example 11.7

The enthalpy of combustion of propane, $C_3H_8$, is $-2219\,kJ\,mol^{-1}$.

What mass of propane must be burned to raise the temperature of $150\,cm^3$ of water by $5\,°C$?

### Solution

Use the equation, $E_h = cm\Delta T$, to calculate the heat required.

$E_h = cm\Delta T = 4.18 \times 0.15 \times 5 = 3.135\,kJ$

Next, we have to form a relationship between energy and mass of propane. This is achieved by recognising that the energy released by 1 mol (44 g) of propane is 2219 kJ.

**energy → mass**

$2219 \rightarrow 44\,g$

$1 \rightarrow \dfrac{44}{2219} = 0.0198\,g$

$3.135 \rightarrow 3.135 \times 0.0198 = \mathbf{0.062\,g}$

## Worked example 11.8

The enthalpy of combustion of methanol is $-726\,kJ\,mol^{-1}$. A burner containing methanol, $CH_3OH$, is used to heat up $400\,cm^3$ of water. What temperature rise would be produced in the water if 0.64 g of methanol were burned completely?

### Solution

In this problem, we cannot use the equation $E_h = cm\Delta T$ immediately as we do not have a temperature rise (we have to calculate the temperature rise). However, the question gives us information about the enthalpy of combustion ($-726\,kJ$) and tells us how much methanol is burned (0.64 g). We can use this to calculate the $E_h$ as follows:

The enthalpy of combustion of $CH_3OH = -726\,kJ\,mol^{-1}$ and 1 mol of $CH_3OH = 32\,g$.

**mass → energy**

$32\,g \rightarrow 726$

$1\,g \rightarrow \dfrac{726}{32} = 22.69\,kJ$

$0.64\,g \rightarrow 0.64 \times 22.69 = 14.52\,kJ$

Use the equation, $E_h = cm\Delta T$, to calculate the temperature change:

$\Delta T = \dfrac{E_h}{cm} = \dfrac{14.52}{4.18 \times 0.4} = \mathbf{8.68\,°C}$

## Questions

23  The enthalpy of combustion of ethanol, $C_2H_5OH$, is $-1367\,kJ\,mol^{-1}$. Calculate the mass of ethanol which, when burned, could raise the temperature of $300\,cm^3$ of water by $10\,°C$.

24  The enthalpy of combustion of propanol, $C_3H_7OH$, is $-2021\,kJ\,mol^{-1}$. Calculate the mass of propanol which, when burned, could raise the temperature of $200\,cm^3$ of water by $13.5\,°C$.

25  A Bunsen burner uses methane, $CH_4$, which has an enthalpy of combustion of $-891\,kJ\,mol^{-1}$. Calculate the maximum temperature rise from heating $500\,cm^3$ of water by burning 0.4 g of methane.

26  The enthalpy of combustion of propane is $-2219\,kJ\,mol^{-1}$. What would be the minimum mass of propane, $C_3H_8$, which would need to be burned completely to bring 5 kg of water to the boil from an initial temperature of $20\,°C$?

27  A camping stove runs on butane, $C_4H_{10}$. If the enthalpy of combustion of butane is $-2878\,kJ\,mol^{-1}$, what mass of butane should be burned to bring 2 kg of water to the boil from an initial temperature of $20\,°C$?

28  The enthalpy of combustion of ethane, $C_2H_6$, is $-1561\,kJ\,mol^{-1}$. 2 g of ethane was burned and the heat released used to raise the temperature of a volume of water by $8\,°C$. Calculate the volume of water heated.

29  The enthalpy of combustion of butane, $C_4H_{10}$, is $-2878\,kJ\,mol^{-1}$. 4 g of butane was burned and the heat released used to raise the temperature of a volume of water by $16\,°C$. Calculate the volume of water heated.

30  The enthalpy of combustion of methane, $CH_4$, is $-891\,kJ\,mol^{-1}$. 0.4 g of methane was used to heat $100\,cm^3$ of water. The highest temperature recorded was $75\,°C$. Calculate the initial temperature of the water.

# 12 Bond enthalpy

The energy changes we calculated in the previous chapter arise from the fact that chemical reactions involve making new substances. In order to do this, reactant bonds must be broken and product bonds must be formed. When this happens, there is an energy change: this is the enthalpy change for the reaction. This chapter will explore calculations which require you to calculate the enthalpy change by considering the energy required to break reactant bonds and the energy released when product bonds are formed.

## Bond breaking and bond making

Consider the reaction of hydrogen with chlorine to form hydrogen chloride:

$$H_2(g) + Cl_2(g) \rightarrow 2HCl(g)$$

This can be represented as shown in Figure 12.1.

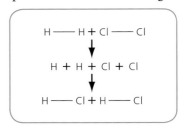

**Figure 12.1** Bond breaking and making in the formation of hydrogen chloride from hydrogen and chlorine

This figure shows that the reaction involves breaking the H–H and Cl–Cl bonds to form H and Cl atoms. These atoms then combine to form two molecules of HCl.

Breaking bonds requires energy (to overcome the attraction between the bonding electrons and the positive nuclei). In other words, breaking bonds is an endothermic process. When new bonds are formed, in this case the bond between H and Cl, energy is released, i.e. bond making is an exothermic process. If you know the energy required to break the bonds and the energy released when the new bonds are made, you can calculate the overall enthalpy change.

## Using bond enthalpies to calculate the enthalpy change for a reaction

Page 10 of the data booklet lists values for the bond enthalpy for selected covalent bonds. For example, the bond enthalpy for H–H is 436 kJ mol$^{-1}$. This means that it takes 436 kJ of energy to break 1 mole of H–H bonds.

Mean bond enthalpies are average values for the energy required to break 1 mol of bonds, taking into account the different types of molecules where these bonds can be found. For example, the value for a C–H bond is listed as a 'mean bond enthalpy' reflecting the fact that a C–H bond can occur in a variety of molecules (alkanes, alcohols, carboxylic acids, etc.). Compare this with a Cl–Cl or H–H bond enthalpy. These bonds cannot occur in any other molecule other than a Cl$_2$ or H$_2$ molecule so the values listed are not average values, they are the exact values.

### Worked example 12.1

**Using bond enthalpies, calculate the enthalpy change for the following reaction:**

$$H_2(g) + Cl_2(g) \rightarrow 2HCl(g)$$

**Solution**

Bond breaking = H–H + Cl–Cl

= 436 + 243

= 679 kJ mol$^{-1}$

Bond making = 2 × (H–Cl)

= 2 × 432

= 864 kJ mol$^{-1}$

The overall enthalpy change = bond making + bond breaking, but you must remember to give the bond making value a negative (–) sign since it is exothermic.

$\Delta H$ = 679 + (–864) = **–185 kJ mol$^{-1}$**

## Worked example 12.2

Using bond enthalpies, calculate the enthalpy change for the following reaction:

$$2H_2(g) + O_2(g) \rightarrow 2H_2O(g)$$

**Solution**

Bond breaking = $2 \times$ (H–H) + (O=O)

$\qquad$ = $2 \times 436 + 498$

$\qquad$ = $1370 \, kJ \, mol^{-1}$

Bond making = $2 \times$ (H–O + H–O)

$\qquad$ = $2 \times (463 + 463)$

$\qquad$ = $1852 \, kJ \, mol^{-1}$

$\Delta H = 1370 + (-1852) = \mathbf{-482 \, kJ \, mol^{-1}}$

## Worked example 12.3

Using bond enthalpies, calculate the enthalpy change for the following reaction:

$$CH_4(g) + Cl_2(g) \rightarrow CH_3Cl(g) + HCl(g)$$

**Solution**

Bond breaking = $4 \times$ (C–H) + (Cl–Cl)

$\qquad$ = $4 \times (412) + 243$

$\qquad$ = $1891 \, kJ \, mol^{-1}$

Bond making = $3 \times$ (C–H) + (C–Cl) + (H–Cl)

$\qquad$ = $3 \times (412) + 338 + 432$

$\qquad$ = $2006 \, kJ \, mol^{-1}$

$\Delta H = 1891 + (-2006) = \mathbf{-115 \, kJ \, mol^{-1}}$

## Worked example 12.4

Experimentally, the bond enthalpies for $H_2(g)$ and $F_2(g)$ were found to be $429 \, kJ \, mol^{-1}$ and $143 \, kJ \, mol^{-1}$, respectively. Using these bond enthalpies and the enthalpy change shown for the following reaction, calculate the experimental value for the bond enthalpy of an H–F bond.

$$H_2(g) + F_2(g) \rightarrow 2HF(g) \quad \Delta H = -510 \, kJ \, mol^{-1}$$

**Solution**

This is a variation on the calculation type. Note that you are **not** using the data booklet values and that you are given the enthalpy change.

Bond breaking = (H–H) + (F–F)

$\qquad$ = $429 + 143$

$\qquad$ = $572 \, kJ \, mol^{-1}$

Bond making = $2 \times$ (H–F)

$\qquad$ = unknown

Enthalpy change = bond breaking + bond making

$\qquad$ $-510 = 572 +$ bond making

$\qquad$ $-510 + (-572) =$ bond making $= -1082$

$2 \times$ (H–F) = $-1082$ so the value of the H–F bond must be **541 kJ mol$^{-1}$**

Note: the bond enthalpy is given as a positive value.

## Questions

**Questions 1–10 are similar to worked examples 12.1–12.3.**

Using bond enthalpies, calculate the enthalpy change for the following reactions.

1  $H_2(g) + F_2(g) \rightarrow 2HF(g)$

2  $H_2(g) + Br_2(g) \rightarrow 2HBr(g)$

3  $H_2(g) + I_2(g) \rightarrow 2HI(g)$

(Note: the structure for $CO_2$ is O=C=O and the structure for $O_2$ is O=O.)

4  $CH_4(g) + 2O_2(g) \rightarrow CO_2(g) + 2H_2O(g)$

5  $C_3H_8(g) + 5O_2(g) \rightarrow 3CO_2(g) + 4H_2O(g)$

6  $2F_2(g) + 2H_2O(g) \rightarrow 4HF(g) + O_2(g)$

7

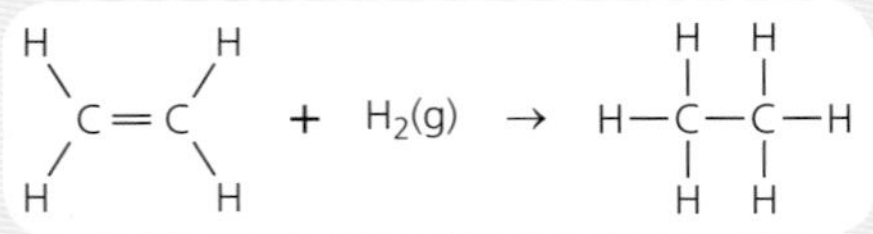

8

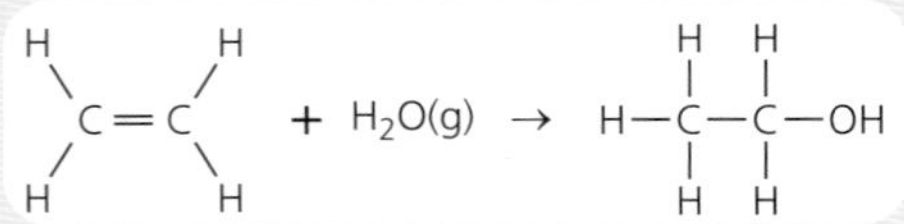

9

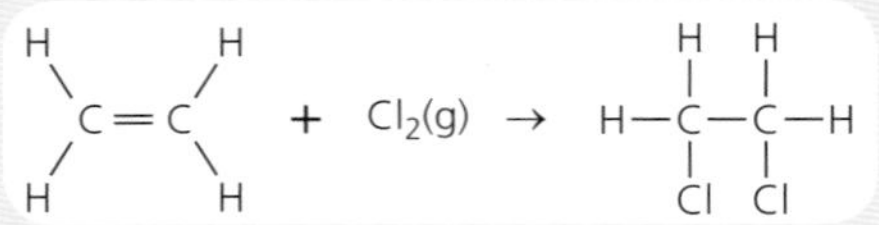

10

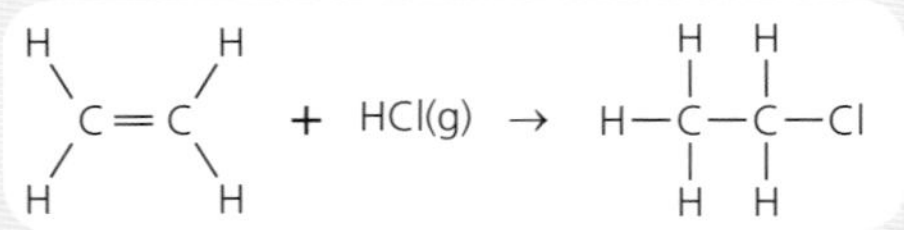

**Questions 11–15 are similar to worked example 12.4.**

11  Use the enthalpy change for the following reaction, and the bond enthalpy of $H_2(g)$ from the data booklet, to calculate the H–S bond enthalpy.

$H_2(g) + S(g) \rightarrow H_2S(g)$ $\Delta H = -17\,kJ\,mol^{-1}$

12  Use the enthalpy change for the following reaction, and the bond enthalpy of $H_2(g)$ from the data booklet, to calculate the H–P bond enthalpy.

$1\frac{1}{2}H_2(g) + P(g) \rightarrow PH_3(g)$ $\Delta H = 5\,kJ\,mol^{-1}$

13  Use the enthalpy change for the following reaction, and the bond enthalpies of $H_2(g)$ and $N_2(g)$ from the data booklet, to calculate the N–H bond enthalpy.

$3H_2(g) + N_2(g) \rightarrow 2NH_3(g)$ $\Delta H = -52\,kJ\,mol^{-1}$

14  Use the enthalpy change for the following reaction, and the bond enthalpies for C≡C, C–H and C–F from the data booklet, to calculate the F–F bond enthalpy.

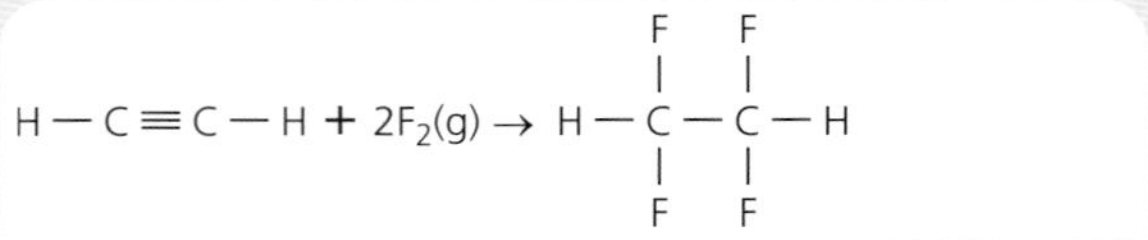

15  Use the enthalpy change for the following reaction, and the bond enthalpies for H–H and C–H from the data booklet, to calculate the C≡C bond enthalpy.

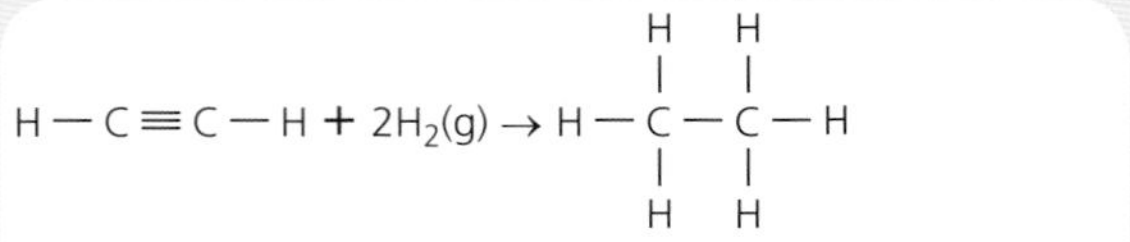

# 13 Hess's Law

Hess's Law of thermochemistry states (in simplified form) that the enthalpy change ($\Delta H$) for a reaction depends only on the enthalpies of the reactants and the products and not on how the reaction is carried out or how many steps are involved in the process.

This allows us to rearrange and combine chemical equations in such a way that we obtain a different equation. We can obtain the $\Delta H$ for the new reaction from the $\Delta H$ values of the original reactions. This idea seems complicated, but the following worked examples should help you understand this concept.

## Worked example 13.1

**Using the enthalpies of combustion of carbon, hydrogen and methane, calculate $\Delta H$ for the following reaction:**

$$C(s) + 2H_2(g) \rightarrow CH_4(g)$$

### Solution: Method 1 (long method)

As the question tells us, the first step is to obtain the relevant equations for the enthalpies of combustion of carbon, hydrogen and methane and the appropriate enthalpy values (from page 10 of the data booklet).

**Enthalpy equations:**

❶ $C(s) + O_2(g) \rightarrow CO_2(g)$ $\Delta H = -394\,kJ\,mol^{-1}$

❷ $H_2(g) + \frac{1}{2}O_2(g) \rightarrow H_2O(l)$ $\Delta H = -286\,kJ\,mol^{-1}$

❸ $CH_4(g) + 2O_2(g) \rightarrow CO_2(g) + 2H_2O(l)$ $\Delta H = -891\,kJ\,mol^{-1}$

The basic technique is to rearrange the equations which we are told to use and to put them into a form which, when added together, will give us the required equation.

The equation we are trying to get to is:

$$C(s) + 2H_2(g) \rightarrow CH_4(g)$$

This is the **target equation**.

The first substance on the left-hand side (LHS) of this equation is $C(s)$ so our first step is to find an equation in the ones we are told to use which contains $C(s)$. The equation labelled ❶ has $C(s)$ on the LHS, so we write it just as it is with its $\Delta H$ value alongside.

❶ $C(s) + O_2(g) \rightarrow CO_2(g)$ $\Delta H = -394\,kJ\,mol^{-1}$

The next substance required is $2H_2(g)$ on the LHS.

Equation ❷ has $H_2(g)$ on the LHS, so this equation and its $\Delta H$ value are doubled. The equation is now represented as $2 \times$ ❷.

$2 \times$ ❷ $2H_2(g) + O_2(g) \rightarrow 2H_2O(l)$ $\Delta H = -572\,kJ\,mol^{-1}$

The last substance we require is $CH_4(g)$ on the right-hand side (RHS).

$CH_4(g)$ is present in equation ❸, but on the LHS, so we **reverse** this equation, change the sign of its $\Delta H$ value and write it as below as $-$❸.

$-$❸ $CO_2(g) + 2H_2O(l) \rightarrow CH_4(g) + 2O_2(g)$ $\Delta H = +891\,kJ\,mol^{-1}$

Collecting these three equations together, we can then cancel out species common to both sides* and add them, and their $\Delta H$ values, together as shown below.

| | | |
|---|---|---|
| ❶ | $C(s) + O_2(g) \rightarrow CO_2(g)$ | $\Delta H = -394\,kJ\,mol^{-1}$ |
| $2 \times$ ❷ | $2H_2(g) + O_2(g) \rightarrow 2H_2O(l)$ | $\Delta H = -572\,kJ\,mol^{-1}$ |
| $-$❸ | $CO_2(g) + 2H_2O(l) \rightarrow CH_4(g) + 2O_2(g)$ | $\Delta H = +891\,kJ\,mol^{-1}$ |
| | $C(s) + 2H_2(g) \rightarrow CH_4(g)$ | $\Delta H = -75\,kJ\,mol^{-1}$ |

*The cancelling took place as follows:

- Two $O_2(g)$, one from the LHS of the first equation and one from the LHS of the second equation, cancel with the $2O_2(g)$ on the RHS of the third equation.

- $2H_2O(l)$ appears on the RHS of the second equation and on the LHS of the third. They are cancelled.

- $CO_2(g)$ appears on the RHS of the first equation and on the LHS of the third equation. They are cancelled.

The equation that we obtain after the adding up and cancelling of the rearranged equations is exactly the one that we require.

**So the calculated $\Delta H$ value of $-75\,kJ\,mol^{-1}$ is the required enthalpy change.**

## Solution: Method 2 (quick method)

The previous solution illustrates how Hess's Law works and allows you to see how the overall equation can be obtained by cancelling the reactants and products from the enthalpy of combustion equations. A quick method involves:

1   identifying the enthalpies of combustion required from the data booklet

2   multiplying each $\Delta H$ by the number of moles in the target equation

3   reversing any enthalpy values where the substance is a product

4   adding the enthalpy values.

So, let's apply this method to the question:

**Using the enthalpies of combustion of carbon, hydrogen and methane, calculate the ΔH for the following reaction:**

$$C(s) + 2H_2(g) \rightarrow CH_4(g)$$

The required enthalpy of combustion equations are:

$1 \times \Delta H$ for C    $+ 2 \times \Delta H$ for $H_2$    $+ 1 \times$ (reverse) $\Delta H$ for $CH_4$

$= -394$        $+ 2 (-286)$        $+ (-1 \times -891)$

$= -75 \text{ kJ mol}^{-1}$

# Writing equations for the enthalpy of combustion

The enthalpy of combustion is defined as **the amount of energy given out when 1 mol of a substance is burned completely in oxygen.**

- When carbon burns completely in oxygen it forms carbon dioxide.

- When hydrogen burns completely in oxygen it forms water.

The enthalpies of combustion of carbon and hydrogen are represented by the equations and $\Delta H$ values shown below.

$$C(s) + O_2(g) \rightarrow CO_2(g) \qquad \Delta H = -394 \text{ kJ mol}^{-1}$$

$$H_2(g) + \frac{1}{2}O_2(g) \rightarrow H_2O(l) \qquad \Delta H = -286 \text{ kJ mol}^{-1}$$

When a compound containing carbon and hydrogen, or carbon, hydrogen and oxygen, burns completely, carbon dioxide and water are formed.

For example, the enthalpy of combustion of methane, $CH_4$, is represented by the equation and $\Delta H$ value shown below.

$$CH_4(g) + 2O_2(g) \rightarrow CO_2(g) + 2H_2O(l)$$
$$\Delta H = -891 \text{ kJ mol}^{-1}$$

## Worked example 13.2

The formation of methanol from its elements is represented by the equation below:

$$C(s) + 2H_2(g) + \tfrac{1}{2}O_2(g) \rightarrow CH_3OH(l)$$

Calculate $\Delta H$ for this reaction using the enthalpies of combustion of carbon, hydrogen and methanol from the data booklet.

### Solution: Method 1 (long method)

The three combustion equations and their $\Delta H$ values, obtained from the data booklet, are written and labelled as ❶, ❷ and ❸ for ease of reference.

❶ $C(s) + O_2(g) \rightarrow CO_2(g)$      $\Delta H = -394\,kJ\,mol^{-1}$

❷ $H_2(g) + \tfrac{1}{2}O_2(g) \rightarrow H_2O(l)$     $\Delta H = -286\,kJ\,mol^{-1}$

❸ $CH_3OH(l) + 1\tfrac{1}{2}O_2(g) \rightarrow CO_2(g) + 2H_2O(l)$

                            $\Delta H = -726\,kJ\,mol^{-1}$

They are then rewritten below, multiplying and/or reversing them to suit the form of the required equation for the formation of methanol. The $\Delta H$ values, multiplied and/or with changed signs where necessary, are written alongside.

| ❶ | $C(s) + O_2(g) \rightarrow CO_2(g)$ | $\Delta H = -394\,kJ\,mol^{-1}$ |
|---|---|---|
| $2 \times$ ❷ | $2H_2(g) + O_2(g) \rightarrow 2H_2O(l)$ | $\Delta H = -572\,kJ\,mol^{-1}$ |
| $-$❸ | $CO_2(g) + 2H_2O(l) \rightarrow CH_3OH(l) + 1\tfrac{1}{2}O_2(g)$ | $\Delta H = +726\,kJ\,mol^{-1}$ |

The rearranged equations and their $\Delta H$ values can then be added up, after cancelling of species common to both sides ($O_2(g)$, $CO_2(g)$ and $H_2O(l)$), to give:

$C(s) + 2H_2(g) + \tfrac{1}{2}O_2(g) \rightarrow CH_3OH(l)$ $\Delta H = -240\,kJ\,mol^{-1}$

The $\Delta H$ for this reaction is, therefore, $-240\,kJ\,mol^{-1}$.

### Solution: Method 2 (quick method)

The required equation is:

$$C(s) + 2H_2(g) + \tfrac{1}{2}O_2(g) \rightarrow CH_3OH(l)$$

The required enthalpy of combustion equations are:

$1 \times \Delta H$ for C   $+ 2 \times \Delta H$ for $H_2$ $+ 1 \times$ (reverse) $\Delta H$ for $CH_3OH$

$= -394$     $+ 2(-286)$     $+ (-1 \times -726)$

$= -240\,kJ\,mol^{-1}$

## Questions

Questions 1–10 supply the necessary equations and enthalpy values. Your task is to use these equations to form the target equation.

**1** The equation for the combustion of methane, $CH_4$, is given below:

$$CH_4(g) + 2O_2(g) \rightarrow CO_2(g) + 2H_2O(l)$$

Calculate the enthalpy of combustion of methane, using the data below:

$C(s) + O_2(g) \rightarrow CO_2(g)$    $\Delta H = -394\,kJ\,mol^{-1}$

$H_2(g) + \tfrac{1}{2}O_2(g) \rightarrow H_2O(l)$    $\Delta H = -286\,kJ\,mol^{-1}$

$C(s) + 2H_2(g) \rightarrow CH_4(g)$    $\Delta H = -75\,kJ\,mol^{-1}$

**2** The equation representing the combustion of ethane, $C_2H_6$, is given below:

$$C_2H_6(g) + 3\tfrac{1}{2}O_2(g) \rightarrow 2CO_2(g) + 3H_2O(l)$$

Calculate the enthalpy of combustion of ethane, $C_2H_6$, using the data below:

$2C(s) + 3H_2(g) \rightarrow C_2H_6(g)$    $\Delta H = -85\,kJ\,mol^{-1}$

$C(s) + O_2(g) \rightarrow CO_2(g)$    $\Delta H = -394\,kJ\,mol^{-1}$

$H_2(g) + \tfrac{1}{2}O_2(g) \rightarrow H_2O(l)$    $\Delta H = -286\,kJ\,mol^{-1}$

**3** The enthalpy of formation of ethane, $C_2H_6$, is represented by the following equation:

$$2C(s) + 3H_2(g) \rightarrow C_2H_6(g)$$

Calculate the enthalpy of formation of ethane using the enthalpies of combustion of hydrogen, carbon and ethane represented by the equations and $\Delta H$ values below:

$C(s) + O_2(g) \rightarrow CO_2(g)$    $\Delta H = -394\,kJ\,mol^{-1}$

$H_2(g) + \tfrac{1}{2}O_2(g) \rightarrow H_2O(l)$    $\Delta H = -286\,kJ\,mol^{-1}$

$C_2H_6(s) + 3\tfrac{1}{2}O_2(g) \rightarrow 2CO_2(g) + 3H_2O(l)$

                 $\Delta H = -1560\,kJ\,mol^{-1}$

4  The enthalpy of combustion of propane, $C_3H_8$, is the $\Delta H$ for the following reaction:

$$C_3H_8(g) + 5O_2(g) \rightarrow 3CO_2(g) + 4H_2O(l)$$

Calculate the enthalpy of combustion of propane using the enthalpy of formation of propane and the enthalpies of combustion of carbon and hydrogen. The equations and $\Delta H$ values for these processes are given below:

$3C(s) + 4H_2(g) \rightarrow C_3H_8(g)$ $\qquad \Delta H = -104\,kJ\,mol^{-1}$

$C(s) + O_2(g) \rightarrow CO_2(g)$ $\qquad \Delta H = -394\,kJ\,mol^{-1}$

$H_2(g) + \frac{1}{2}O_2(g) \rightarrow H_2O(l)$ $\qquad \Delta H = -286\,kJ\,mol^{-1}$

5  The formation of butane, $C_4H_{10}$, is represented by the equation:

$$4C(s) + 5H_2(g) \rightarrow C_4H_{10}(g)$$

Calculate the enthalpy of formation of butane using the enthalpies of combustion of butane, hydrogen and carbon represented by the equations and $\Delta H$ values below:

$C_4H_{10}(g) + 6\frac{1}{2}O_2(g) \rightarrow 4CO_2(g) + 5H_2O(l)$

$\qquad\qquad\qquad\qquad \Delta H = -2877\,kJ\,mol^{-1}$

$H_2(g) + \frac{1}{2}O_2(g) \rightarrow H_2O(l)$ $\qquad \Delta H = -286\,kJ\,mol^{-1}$

$C(s) + O_2(g) \rightarrow CO_2(g)$ $\qquad \Delta H = -394\,kJ\,mol^{-1}$

6  The equation for the combustion of ethanol, $C_2H_5OH$, is:

$$C_2H_5OH(l) + 3O_2(g) \rightarrow 2CO_2(g) + 3H_2O(l)$$

Calculate the enthalpy of combustion of ethanol using the enthalpies of combustion of carbon and hydrogen, and the enthalpy of formation of ethanol represented by the equations and $\Delta H$ values below:

$C(s) + O_2(g) \rightarrow CO_2(g)$ $\qquad \Delta H = -394\,kJ\,mol^{-1}$

$H_2(g) + \frac{1}{2}O_2(g) \rightarrow H_2O(l)$ $\qquad \Delta H = -286\,kJ\,mol^{-1}$

$2C(s) + 3H_2(g) + \frac{1}{2}O_2(g) \rightarrow C_2H_5OH(l)$

$\qquad\qquad\qquad\qquad \Delta H = -278\,kJ\,mol^{-1}$

7  The formation of ethanoic acid, $CH_3COOH$, is represented by the equation:

$$2C(s) + 2H_2(g) + O_2(g) \rightarrow CH_3COOH(l)$$

Calculate the enthalpy of formation of ethanoic acid using the enthalpies of combustion of carbon,

hydrogen and ethanoic acid represented by the equations and $\Delta H$ values below:

$C(s) + O_2(g) \rightarrow CO_2(g)$ $\qquad \Delta H = -394\,kJ\,mol^{-1}$

$H_2(g) + \frac{1}{2}O_2(g) \rightarrow H_2O(l)$ $\qquad \Delta H = -286\,kJ\,mol^{-1}$

$CH_3COOH(l) + 2O_2(g) \rightarrow 2CO_2(g) + 2H_2O(l)$

$\qquad\qquad\qquad\qquad \Delta H = -876\,kJ\,mol^{-1}$

8  The formation of propan-1-ol, $C_3H_7OH$, is represented by the equation:

$$3C(s) + 4H_2(g) + \frac{1}{2}O_2(g) \rightarrow C_3H_7OH(l)$$

Calculate the enthalpy of formation of propan-1-ol using the enthalpies of combustion of carbon, hydrogen and propan-1-ol represented by the equations and $\Delta H$ values:

$C(s) + O_2(g) \rightarrow CO_2(g)$ $\qquad \Delta H = -394\,kJ\,mol^{-1}$

$H_2(g) + \frac{1}{2}O_2(g) \rightarrow H_2O(l)$ $\qquad \Delta H = -286\,kJ\,mol^{-1}$

$C_3H_7OH(l) + 4\frac{1}{2}O_2(g) \rightarrow 3CO_2(g) + 4H_2O(l)$

$\qquad\qquad\qquad\qquad \Delta H = -2020\,kJ\,mol^{-1}$

9  The combustion of benzene, $C_6H_6$, is represented by the equation:

$$C_6H_6(l) + 7\frac{1}{2}O_2(g) \rightarrow 6CO_2(g) + 3H_2O(l)$$

Calculate the enthalpy of combustion of benzene using the enthalpies of combustion of carbon and hydrogen and the enthalpy of formation of benzene represented by the equations and $\Delta H$ values:

$C(s) + O_2(g) \rightarrow CO_2(g)$ $\qquad \Delta H = -394\,kJ\,mol^{-1}$

$H_2(g) + \frac{1}{2}O_2(g) \rightarrow H_2O(l)$ $\qquad \Delta H = -286\,kJ\,mol^{-1}$

$6C(s) + 3H_2(g) \rightarrow C_6H_6(l)$ $\qquad \Delta H = +49\,kJ\,mol^{-1}$

10 The formation of ethyne, $C_2H_2$, is represented by the equation:

$$2C(s) + H_2(g) \rightarrow C_2H_2(g)$$

Calculate the enthalpy of formation of ethyne using the enthalpies of combustion of carbon, hydrogen and ethyne represented by the equations and $\Delta H$ values below:

$C(s) + O_2(g) \rightarrow CO_2(g)$ $\qquad \Delta H = -394\,kJ\,mol^{-1}$

$H_2(g) + \frac{1}{2}O_2(g) \rightarrow H_2O(l)$ $\qquad \Delta H = -286\,kJ\,mol^{-1}$

$C_2H_2(g) + 2\frac{1}{2}O_2(g) \rightarrow 2CO_2(g) + H_2O(l)$

$\qquad\qquad\qquad\qquad \Delta H = -1300\,kJ\,mol^{-1}$

**Questions 11–20 require you to write enthalpy of combustion equations and obtain the enthalpy values from the data booklet.**

11 The formation of methanoic acid, HCOOH, is represented by the following equation:

$$C(s) + H_2(g) + O_2(g) \rightarrow HCOOH(l)$$

Calculate the $\Delta H$ for the above reaction using the enthalpies of combustion of carbon, hydrogen and methanoic acid.

12 The reaction of ethanol with oxygen to form ethanoic acid and water is represented by the equation:

$$C_2H_5OH(l) + O_2(g) \rightarrow CH_3COOH(l) + H_2O(l)$$

Use the enthalpies of combustion of ethanol and ethanoic acid to calculate the $\Delta H$ for the above reaction.

13 The equation below represents the hydrogenation of ethene to ethane:

$$C_2H_4(g) + H_2(g) \rightarrow C_2H_6(g)$$

Use the enthalpies of combustion of ethene, hydrogen and ethane to calculate the $\Delta H$ for the above reaction.

14 The fermentation of glucose, $C_6H_{12}O_6$, to ethanol and carbon dioxide can be represented by the equation:

$$C_6H_{12}O_6(s) \rightarrow 2C_2H_5OH(l) + 2CO_2(g)$$

Calculate the $\Delta H$ for this reaction using the enthalpies of combustion of ethanol and glucose. (The enthalpy of combustion of glucose is $-2813\,kJ\,mol^{-1}$.)

15 Use the enthalpies of combustion of ethyne, $C_2H_2$, ethane, $C_2H_6$, and hydrogen to calculate the $\Delta H$ for the complete hydrogenation of ethyne as given by the equation:

$$C_2H_2(s) + 2H_2(g) \rightarrow C_2H_6(g)$$

16 The enthalpy of formation of diborane, $B_2H_6$, is the $\Delta H$ for the following reaction:

$$2B(s) + 3H_2(g) \rightarrow B_2H_6(g)$$

Calculate the enthalpy of formation of diborane using the enthalpy of combustion of hydrogen and the equations and $\Delta H$ values for the enthalpy of formation of boron oxide and the enthalpy of combustion of diborane noted below:

$$2B(s) + 1\tfrac{1}{2}O_2(g) \rightarrow B_2O_3(s) \qquad \Delta H = -612\,kJ\,mol^{-1}$$

$$B_2H_6(g) + 3O_2(g) \rightarrow B_2O_3(s) + 3H_2O(l)$$
$$\Delta H = -1058\,kJ\,mol^{-1}$$

17 Use the enthalpies of combustion of methane, hydrogen and ethyne, $C_2H_2$, to obtain the $\Delta H$ for the reaction represented by the equation:

$$2CH_4(g) \rightarrow C_2H_2(g) + 3H_2(g)$$

18 $Na(s) + \tfrac{1}{2}Cl_2(g) \rightarrow Na^+(g) + Cl^-(g)$
$$\Delta H = +365\,kJ\,mol^{-1}$$

$Na(s) + \tfrac{1}{2}Cl_2(g) \rightarrow Na^+Cl^-(s) \qquad \Delta H = -411\,kJ\,mol^{-1}$

Use the above information to calculate the $\Delta H$ for the reaction:

$$Na^+Cl^-(s) \rightarrow Na^+(g) + Cl^-(g)$$

19 $Cu(s) \rightarrow Cu^{2+}(aq) + 2e^- \qquad \Delta H = +795\,kJ\,mol^{-1}$

$Cu(s) \rightarrow Cu^+(aq) + e^- \qquad \Delta H = +602\,kJ\,mol^{-1}$

Use the above information to calculate the $\Delta H$ for the reaction:

$$2Cu^+(aq) \rightarrow Cu^{2+}(aq) + Cu(s)$$

20 The enthalpies of formation of methylhydrazine, $CH_3NHNH_2$, and dinitrogen tetroxide, $N_2O_4$, are represented by the equations and $\Delta H$ values below:

$$C(s) + 3H_2(g) + N_2(g) \rightarrow CH_3NHNH_2(l)$$
$$\Delta H = +53\,kJ\,mol^{-1}$$

$$N_2(g) + 2O_2(g) \rightarrow N_2O_4(g) \qquad \Delta H = -20\,kJ\,mol^{-1}$$

Use this information, and the enthalpies of combustion of carbon and hydrogen, to obtain the $\Delta H$ for the equation:

$$4CH_3NHNH_2(l) + 5N_2O_4(g) \rightarrow 4CO_2(g) + 12H_2O(l) + 9N_2(g)$$

## 14 Volumetric calculations

This chapter explores calculations from equations where the quantity of one or both of the substances referred to is expressed in terms of the concentration and volume of a solution. This is usually encountered in the Higher course in titration questions involving acids and bases or in redox reactions.

## Volumetric analysis: simple acid/base titrations

### Worked example 14.1

**Calculate the volume of 0.5 mol l⁻¹ sodium hydroxide solution that will neutralise 40 cm³ of 0.2 mol l⁻¹ sulfuric acid.**

$$2NaOH + H_2SO_4 \rightarrow Na_2SO_4 + 2H_2O$$

**Solution**

The method used to solve these problems is identical to solving any calculation from an equation, except that the chemicals reacting are solutions. This means that to calculate the number of moles reacting, you use the expression

**moles = concentration × volume**

**Step 1: Calculation of 'known' moles**

In this case, we are told the volume and concentration of the sulfuric acid. This enables us to calculate the number of moles.

moles = concentration × volume (in litres)
$\quad$ = 0.04 × 0.2
$\quad$ = 0.008 mol (of sulfuric acid)

**Step 2: Mole statement to calculate the 'unknown' moles**

1 mol of $H_2SO_4$ reacts with 2 mol of NaOH

Therefore, 0.008 mol of $H_2SO_4$ reacts with 0.016 mol of NaOH

**Step 3: Finishing off**

The problem asks for the volume of NaOH solution. We are told in the problem that its concentration is 0.5 mol l⁻¹ and we have just calculated that we need 0.016 mol of NaOH.

$$volume = \frac{moles}{concentration} = \frac{0.016}{0.5} = 0.032 \text{ litres} = \mathbf{32\ cm^3}$$

### Worked example 14.2

**A 40 cm³ sample of sodium chloride solution is reacted with an excess of silver(I) nitrate solution, $AgNO_3$. The precipitate of silver(I) chloride formed is filtered, dried and weighed; its mass is found to be 7.17 g. Calculate the concentration of the sodium chloride solution, in mol l⁻¹.**

$$AgNO_3 + NaCl \rightarrow AgCl + NaNO_3$$

**Solution**

**Step 1: Calculation of 'known' moles**

In this case, we are told the mass of the silver(I) chloride precipitate, which allows us to calculate the number of moles of AgCl:

$$moles = \frac{mass}{gfm} = \frac{7.17}{143.4} = 0.05 \text{ mol of AgCl}$$

**Step 2: Mole statement to calculate the 'unknown' moles**

1 mol of NaCl will produce 1 mol of AgCl precipitate

So, 0.05 mol of AgCl would have come from 0.05 mol of NaCl.

**Step 3: Finishing off**

The problem asks for the concentration of the NaCl solution, and tells us that its volume is 40 cm³ (0.04 litres).

$$volume = \frac{moles}{concentration} = \frac{0.05}{0.04} = \mathbf{1.25\ litres}$$

## Questions

These problems are of the type illustrated in worked examples 14.1 and 14.2 involving the concentrations and volumes of solutions.

1 $50\,cm^3$ of $0.4\,mol\,l^{-1}$ potassium hydroxide solution neutralised $20\,cm^3$ of sulfuric acid.

Calculate the concentration of the sulfuric acid.

$2KOH + H_2SO_4 \rightarrow K_2SO_4 + 2H_2O$

2 $20\,cm^3$ of $0.2\,mol\,l^{-1}$ sulfuric acid neutralised a volume of $0.5\,mol\,l^{-1}$ sodium hydroxide solution. Calculate the volume of the sodium hydroxide solution.

$2NaOH + H_2SO_4 \rightarrow Na_2SO_4 + 2H_2O$

3 $200\,cm^3$ of $0.5\,mol\,l^{-1}$ potassium hydroxide solution was reacted with an excess of carbon dioxide. What mass of potassium carbonate would be obtained upon complete evaporation of the water from the resulting solution?

$2KOH + CO_2 \rightarrow K_2CO_3 + H_2O$

4 $1.96\,g$ of pure phosphoric acid, $H_3PO_4$, was dissolved in water and the solution neutralised by a quantity of $0.2\,mol\,l^{-1}$ sodium hydroxide solution. Calculate the volume of the sodium hydroxide solution.

$H_3PO_4 + 3NaOH \rightarrow Na_3PO_4 + 3H_2O$

5 $6.8\,g$ of ammonia gas was used to neutralise $250\,cm^3$ of sulfuric acid. Calculate the concentration of the sulfuric acid.

$2NH_3 + H_2SO_4 \rightarrow (NH_4)_2SO_4$

6 A solution of phthalic acid, $C_6H_4(COOH)_2$, was neutralised by $25\,cm^3$ of a $2\,mol\,l^{-1}$ solution of sodium hydroxide. Calculate the mass of phthalic acid that was dissolved in the acid solution.

$C_6H_4(COOH)_2 + 2NaOH \rightarrow C_6H_4(COONa)_2 + 2H_2O$

7 A $20\,cm^3$ sample of vinegar (dilute ethanoic acid, $CH_3COOH$) was titrated with standard $0.1\,mol\,l^{-1}$ sodium hydroxide solution. Exactly $36.4\,cm^3$ of the alkali was found to neutralise the acid sample. Calculate the concentration of the ethanoic acid.

$CH_3COOH + NaOH \rightarrow CH_3COONa + H_2O$

8 $12.5\,cm^3$ of $0.4\,mol\,l^{-1}$ solution of sulfuric acid was poured into a solution containing an excess of barium chloride. A precipitate of barium sulfate was formed which was filtered, washed, dried and weighed. What mass of precipitate should theoretically be obtained?

$H_2SO_4 + BaCl_2 \rightarrow BaSO_4 + 2HCl$

9 A solution of oxalic acid, $(COOH)_2$, was standardised by taking a $25\,cm^3$ sample of it and titrating it against $0.2\,mol\,l^{-1}$ potassium hydroxide solution. Neutralisation was obtained by the addition of $23.5\,cm^3$ of the potassium hydroxide. Calculate the concentration, in $mol\,l^{-1}$, of the oxalic acid.

$2KOH + (COOH)_2 \rightarrow (COOK)_2 + 2H_2O$

10 $20\,cm^3$ of dilute sulfuric acid was reacted with an excess of lead(II) nitrate, causing $1.2132\,g$ of lead(II) sulfate to be precipitated. Calculate the concentration, in $mol\,l^{-1}$, of the acid.

$Pb(NO_3)_2 + H_2SO_4 \rightarrow PbSO_4 + 2HNO_3$

# Volumetric analysis: titrations involving dilutions

## Worked example 14.3

A sample of sulfuric acid is to be analysed by titration with sodium carbonate solution.

$$Na_2CO_3 + H_2SO_4 \rightarrow Na_2SO_4 + CO_2 + H_2O$$

A 25 cm$^3$ sample of it is pipetted into a 250 cm$^3$ standard flask and the solution is made up to the mark. A 25 cm$^3$ sample of this diluted solution is titrated with standard 0.2 mol l$^{-1}$ sodium carbonate solution; it is found that 15 cm$^3$ of the sodium carbonate solution is required to neutralise the acid sample. What concentration was the original (undiluted) acid?

### Solution

In problems which involve samples being diluted, it is difficult to generalise about the best method. However, dilutions are usually made in multiples of 10; when this is the case the problem can usually be solved without too much difficulty. In this problem, a 25 cm$^3$ sample is diluted to 250 cm$^3$; in other words *it has been diluted to $\frac{1}{10}$th of its original concentration*. With that in mind, we can first work out the concentration of the *diluted* acid, and simply multiply this value by 10 to get the original concentration.

### Step 1: Calculation of 'known' moles

In this case, we are told the volume and concentration of the sodium carbonate solution. This enables us to calculate the number of moles.

moles = concentration × volume (in litres)
    = 0.2 × 0.015
    = 0.003 mol of Na$_2$CO$_3$

### Step 2: Mole statement to calculate the 'unknown' moles

1 mol of Na$_2$CO$_3$ neutralises 1 mol of H$_2$SO$_4$

0.003 mol of Na$_2$CO$_3$ will neutralise 0.003 mol of H$_2$SO$_4$

### Step 3: Check for dilution

In this case, the sodium carbonate solution is reacting with 25 cm$^3$ of the acid solution.

0.003 mol of sulfuric acid is present in 25 cm$^3$ solution.

0.03 mol of sulfuric acid must be present in 250 cm$^3$ solution.

### Step 4: Finishing off

The problem asks for the concentration of the sulfuric acid, and tells us that its original, undiluted volume was 25 cm$^3$ (0.025 litres).

$$concentration = \frac{moles}{volume}$$
$$= \frac{0.03}{0.025}$$
$$= 1.2 \text{ mol l}^{-1}$$

## Questions

These problems involve the dilution of solutions as in worked example 14.3.

11  A 25 cm³ sample of sulfuric acid of unknown concentration was pipetted into a 250 cm³ standard flask and made up to the mark with water. A 25 cm³ sample of this solution was titrated with 0.2 mol l⁻¹ sodium hydroxide solution. 50 cm³ of sodium hydroxide solution was required to neutralise the acid sample. Calculate the concentration of the original 25 cm³ sample of sulfuric acid.

$$2NaOH + H_2SO_4 \rightarrow Na_2SO_4 + 2H_2O$$

12  30 g of pure ethanoic acid, $CH_3COOH$, was diluted with water to make up a 250 cm³ standard solution. A 10 cm³ sample of this diluted acid was titrated with 0.2 mol l⁻¹ potassium hydroxide solution. Calculate the volume of the potassium hydroxide required to neutralise the 10 cm³ sample.

$$CH_3COOH + KOH \rightarrow CH_3COOK + H_2O$$

13  19.62 g of pure sulfuric acid was weighed and made up to 1 litre with water in a standard flask. A 40 cm³ sample of this diluted solution was found to be neutralised exactly by 25 cm³ of sodium carbonate solution. Calculate the concentration of the sodium carbonate solution.

$$H_2SO_4 + Na_2CO_3 \rightarrow Na_2SO_4 + CO_2 + H_2O$$

14  Car batteries contain sulfuric acid. A 10 cm³ sample of battery acid was pipetted into a 1 litre standard flask and made up to the mark with water. A 50 cm³ sample of the new diluted acid was titrated against standard 0.2 mol l⁻¹ sodium hydroxide solution. It was found that 12.5 cm³ of the sodium hydroxide was required to neutralise the acid sample. Calculate the concentration of the battery acid.

$$2NaOH + H_2SO_4 \rightarrow Na_2SO_4 + 2H_2O$$

15  A standard 1 mol l⁻¹ solution of sodium carbonate was prepared. A 20 cm³ sample of this sodium carbonate solution was then withdrawn and added to a 500 cm³ standard flask which was made up to the mark with water. A 20 cm³ sample of this *diluted solution* was titrated against hydrochloric acid of unknown concentration. 40 cm³ of acid was required to neutralise the sample.

$$2HCl + Na_2CO_3 \rightarrow 2NaCl + CO_2 + H_2O$$

a)  How many moles of sodium carbonate were present in the first 20 cm³ sample?

b)  Calculate the concentration of the hydrochloric acid.

# Volumetric analysis: redox titrations

In acid–alkali reactions, the concentration of one solution can be obtained by titration with another of known concentration, using an indicator which changes colour at the 'end-point' – the point where one substance has *just* neutralised the other.

Redox reactions can also be used to find the concentration of an 'unknown' solution. Sometimes the redox reaction is 'self-indicating', meaning that one of the reactants changes colour sharply at the end-point. Common reactants of this type include the permanganate ion, $MnO_4^-(aq)$ and the dichromate ion,

$Cr_2O_7^{2-}(aq)$, which, in acid solution, can undergo the following reductions, with the colour changes shown. (These ion–electron equations are included on page 13 of the SQA data booklet.)

$$MnO_4^-(aq) + 8H^+(aq) + 5e^- \rightarrow Mn^{2+}(aq) + 14H_2O(l)$$

PURPLE                                    COLOURLESS

$$Cr_2O_7^{2-}(aq) + 14H^+(aq) + 6e^- \rightarrow Cr^{3+}(aq) + 7H_2O(l)$$

ORANGE                                    GREEN

Other redox reactions will require the presence of an added indicator which will change colour at the end-point. A common example is starch which forms a purple-black colour in the presence of iodine, $I_2$.

## Worked example 14.4

20 cm³ of a 0.2 mol l⁻¹ solution of iodide ions was titrated with a 0.064 mol l⁻¹ solution of permanganate ions, in acid solution.

The ion–electron equations below represent the oxidation and reduction reactions taking place when permanganate ions, $MnO_4^-$, in acid solution react with iodide ions.

$$MnO_4^- + 8H^+ + 5e^- \rightarrow Mn^{2+} + 4H_2O$$
$$2I^- \rightarrow I_2 + 2e^-$$

What volume of the permanganate solution will be required to exactly react with all the iodide ions?

### Solution

#### Step 1: Obtain a balanced redox equation

This problem is actually not different from those done previously, except that the balanced equation for the overall redox reaction has not been given. Using the ion–electron equations given, it can be seen that the reduction reaction involves the gain of five electrons, while the oxidation reaction involves the loss of two electrons. Since the number of electrons being gained must be the same as that being lost, the balanced equation is obtained by multiplying the above equations as below:

2 × RED: 2 × ($MnO_4^- + 8H^+ + 5e^- \rightarrow Mn^{2+} + 4H_2O$)

5 × OX: 5 × ($2I^- \rightarrow I_2 + 2e^-$)

If the reduction is multiplied throughout by 2, we now have a **gain of 10 electrons**. The multiplication of the oxidation by 5 gives a **loss of 10 electrons**.

That is, the two 'half-reactions' are now balanced. The full balanced equation can be obtained by multiplying out the reduction and oxidation reactions, adding them together and cancelling out species common to each side to give:

$$2MnO_4^- + 16H^+ + 10I^- \rightarrow 2Mn^{2+} + 8H_2O + 5I_2$$

This tells us that:

2 mol of $MnO_4^-$ reacts with 10 mol of $I^-$.

Or, more simply, 1 mol of $MnO_4^-$ reacts with 5 mol of $I^-$.

#### Step 2: Calculation of 'known' moles

In this case we are told the volume and concentration of the iodide solution. This enables us to calculate the number of moles.

moles = concentration × volume (in litres)

$\qquad$ = 0.2 × 0.02

$\qquad$ = 0.0004 mol of iodide

#### Step 3: Calculation of 'unknown' moles

From the mole statement:

1 mol of $MnO_4^-$ reacts with 5 mol of $I^-$

Reversing to put $MnO_4^-$ on the right-hand side, we have:

5 mol of $I^-$ reacts with 1 mol of $MnO_4^-$.

1 mol of $I^-$ reacts with $\frac{1}{5}$ mol of $MnO_4^-$.

0.004 mol of $I^-$ reacts with $\frac{0.004}{5}$ mol of $MnO_4^-$
= 0.0008 mol of $MnO_4^-$.

#### Step 5: Finishing off

volume (in litres) = $\dfrac{\text{number of moles}}{\text{concentration}}$

$\qquad$ = $\dfrac{0.0008}{0.064}$

$\qquad$ = **0.0125 l (12.5 cm³)**

## Questions

These questions involve redox reactions of the type illustrated in worked example 14.4.

**16** $Cr_2O_7^{2-} + 14H^+ + 6e^- \rightarrow 2Cr^{3+} + 7H_2O$

$SO_3^{2-} + H_2O \rightarrow SO_4^{2-} + 2H^+ + 2e^-$

The above ion–electron equations represent the reduction and oxidation reactions which take place when a solution of dichromate ions, $Cr_2O_7^{2-}$, in acid solution reacts with sulfite ions, $SO_3^{2-}$.

What volume of a $0.05\,mol\,l^{-1}$ solution of dichromate ions would react with $30\,cm^3$ of a $0.25\,mol\,l^{-1}$ solution of sulfite ions?

**17** $MnO_4^- + 8H^+ + 5e^- \rightarrow Mn^{2+} + 4H_2O$

$2Cl^- \rightarrow Cl_2 + 2e^-$

What volume of a $0.24\,mol\,l^{-1}$ solution of acidified permanganate ions would exactly react with $120\,cm^3$ of a $0.16\,mol\,l^{-1}$ solution of chloride ions?

**18** $I_2 + 2e^- \rightarrow 2I^-$

$2S_2O_3^{2-} \rightarrow S_4O_6^{2-} + 2e^-$

$25\,cm^3$ of a $0.05\,mol\,l^{-1}$ solution of thiosulfate ions, $S_2O_3^{2-}$, reacts exactly with $10\,cm^3$ of a solution of $I_2$. What concentration is the $I_2$ solution?

**19** $Fe^{3+} + e^- \rightarrow Fe^{2+}$

$SO_3^{2-} + H_2O \rightarrow SO_4^{2-} + 2H^+ + 2e^-$

$25\,cm^3$ of a solution containing $0.016\,mol\,l^{-1}$ sulfite ions is titrated with $0.02\,mol\,l^{-1}$ $Fe^{3+}$ solution. What volume of the $Fe^{3+}$ solution would be required to obtain the exact end-point of this titration?

**20** $2BrO_3^- + 12H^+ + 10e^- \rightarrow Br_2 + 6H_2O$

$2I^- \rightarrow I_2 + 2e^-$

$32\,cm^3$ of a $0.0125\,mol\,l^{-1}$ solution of bromate ions, $BrO_3^-$, exactly oxidises $25\,cm^3$ of a solution of iodide ions. What is the concentration of the iodide ion solution?

**21** $2HClO + 2H^+ + 2e^- \rightarrow Cl_2 + 2H_2O$

$2I^- \rightarrow I_2 + 2e^-$

$25\,cm^3$ of a $0.02\,mol\,l^{-1}$ solution of hypochlorous acid, HClO, is exactly reduced by $10\,cm^3$ of a solution of iodide ions. What is the concentration of the iodide ion solution?

**22** $Fe^{3+} + e^- \rightarrow Fe^{2+}$

$SO_3^{2-} + H_2O \rightarrow SO_4^{2-} + 2H^+ + 2e^-$

$12.8\,cm^3$ of a $0.12\,mol\,l^{-1}$ solution of sulfite ions is exactly oxidised by $7.68\,cm^3$ of a solution of iron(III) ions. What is the concentration of the iron(III) ions present?

**23** $Cr_2O_7^{2-} + 14H^+ + 6e^- \rightarrow Cr^{3+} + 7H_2O$

$Fe^{2+} \rightarrow Fe^{3+} + e^-$

$40\,cm^3$ of a solution containing acidified dichromate ions with a concentration of $0.015\,mol\,l^{-1}$ is titrated with $0.2\,mol\,l^{-1}$ $Fe^{2+}$ solution. What volume of the $Fe^{2+}$ solution would be required to obtain the exact end-point of this titration?

**24** $(COOH)_2 \rightarrow 2CO_2 + 2H^+ + 2e^-$

$MnO_4^- + 8H^+ + 5e^- \rightarrow Mn^{2+} + 4H_2O$

A $0.01\,mol\,l^{-1}$ solution of oxalic acid, $(COOH)_2$, is titrated against $16\,cm^3$ of an acidified solution containing $0.005\,mol\,l^{-1}$ permanganate ions until the end-point is reached. What volume of oxalic acid solution must have reacted?

**25** $H_2O_2 + 2H^+ + 2e^- \rightarrow 2H_2O$

$Fe^{2+} \rightarrow Fe^{3+} + e^-$

$10\,cm^3$ of a $2.4\,mol\,l^{-1}$ solution of hydrogen peroxide, $H_2O_2$, is titrated with a solution containing $1.25\,mol\,l^{-1}$ of $Fe^{2+}$. What volume of the $Fe^{2+}$ solution will be required to react exactly with the hydrogen peroxide?

# 15 Using numeracy in Higher Chemistry

Being able to 'scale up' or 'scale down' is a useful skill for chemists. Several calculations covered in this book have involved using proportion and have used this scaling technique. An application of this is using proportion to calculate costs, since chemists must be able to cost their experiments in an attempt to make maximum profit when new products are made.

This chapter will give further practise using proportion.

## Using simple proportion

The following worked examples use a common layout. In all examples, a relationship is formed between two quantities with the unknown (whatever you are being asked to calculate) placed on the right-hand side of the arrow.

### Worked example 15.1

The enthalpy of combustion of ethane, $C_2H_6$, is $-1561\,kJ\,mol^{-1}$.

Calculate the energy released when 80 g of ethane is burned.

**Solution**

The relationship is between mass and energy. We are told the energy released by 1 mol, allowing us to calculate the mass of ethane burned (since 1 mol = 30 g).

**Step 1: State a relationship**

Since we are asked to calculate energy, this goes on the right-hand side:

mass → energy

$30\,g \rightarrow -1561$

**Step 2: Scale to 1**

To change 30 g into 1 g, we need to divide 30 by 30. Whatever we do to the left-hand side, we must do to the right-hand side:

$$\frac{30}{30} \rightarrow -\frac{1561}{30}$$

$1\,g \rightarrow -52.03$

**Step 3: Calculate for the quantity you are asked for**

$80\,g \rightarrow 80 \times -52.03 = -4162.6\,kJ\,mol^{-1}$

### Worked example 15.2

In an experiment, 10 g of ethene was burned and released 400 kJ of energy.

Calculate the enthalpy of combustion of ethene (1 mol of ethene = 28 g).

**Solution**

mass → energy

$10\,g \rightarrow 400\,kJ$

$1\,g \rightarrow \dfrac{400}{10} = 40\,kJ$

$28\,g \rightarrow 1120\,kJ$

The enthalpy of combustion of ethene is $-1120\,kJ\,mol^{-1}$.

### Worked example 15.3

The enthalpy of combustion of methane (1 mol = 16 g) is $-891\,kJ\,mol^{-1}$.

Calculate the mass of methane that should be burned to release 100 kJ of energy.

**Solution**

energy → mass

$891\,kJ \rightarrow 16\,g$

$1\,kJ \rightarrow \dfrac{16}{891} = 0.018\,g$

$100\,kJ \rightarrow 100 \times 0.018 = 1.80\,g$

## Worked example 15.4

The cost of 100 g of NaOH(s) is £15.40.

Calculate the cost of 328 g of NaOH(s).

Solution

mass → cost

$100\,g \rightarrow £15.40$

$1\,g \rightarrow \dfrac{15.40}{100} = £0.154$

$328\,g \rightarrow 328 \times £0.154 = \textbf{£50.51}$

## Worked example 15.5

11 g of $CO_2$ contains $1.505 \times 10^{23}$ $CO_2$ molecules.

Calculate the number of molecules present in 100 g of $CO_2$.

Solution

mass → number of molecules

$11\,g \rightarrow 1.505 \times 10^{23}$

$1\,g \rightarrow 1.368 \times 10^{22}$

$100\,g \rightarrow \textbf{1.368} \times \textbf{10}^{\textbf{24}}$

## Worked example 15.6

16 g of $CH_4$ contains $6.02 \times 10^{23}$ $CH_4$ molecules.

Calculate the mass of 12 $CH_4$ molecules.

Solution

number of molecules → mass

$6.02 \times 10^{23}$ molecules $\rightarrow 16\,g$

$1$ molecule $\rightarrow \dfrac{16}{6.02 \times 10^{23}}$

$= 2.66 \times 10^{-23}\,g$

$12$ molecules $\rightarrow 12 \times 2.66 \times 10^{-23}\,g$

$= \textbf{3.19} \times \textbf{10}^{\textbf{-22}}\,\textbf{g}$

# Using proportion: more challenging examples

## Worked example 15.7

An egg contains 3.1 µg of vitamin D. This is equivalent to 31% of the recommended daily allowance (RDA) of vitamin D. Calculate the RDA of vitamin D and, hence, calculate the number of eggs required to supply the RDA.

Solution

Step 1: State a relationship between % and mass to allow you to calculate the RDA

percentage → mass

$31\% \rightarrow 3.1\,µg$

$1\% \rightarrow \dfrac{3.1}{31} = 0.1\,µg$

$100\% \rightarrow 10\,µg$

The RDA of vitamin D is **10 µg**.

Step 2: Calculate the number of eggs that could supply 10 µg

mass → number of eggs

$3.1\,µg \rightarrow 1$ egg

$1\,µg \rightarrow \dfrac{1}{3.1} = 0.323$ eggs

$10\,µg \rightarrow 10 \times 0.323 = \textbf{3.23 eggs}$

Step 2 is essentially obtained by $\dfrac{10}{3.1} = 3.23$

## Worked example 15.8

Dogs suffering from arthritis can be treated for 10 days by supplying one dose per day of the treatment which must be supplied at 5 mg per kg.

Calculate the cost of treating a 34 kg dog for 10 days given that 200 mg of the treatment costs £1.25.

### Solution

mass of dog → mass of treatment required

$$1 \text{ kg} \rightarrow 5 \text{ mg}$$

$$34 \text{ kg} \rightarrow 34 \times 5 = 170 \text{ mg}$$

Therefore, 10 days at 170 mg per day = 1700 mg.

mass → cost

$$200 \text{ mg} \rightarrow £1.25$$

$$1 \text{ mg} \rightarrow \frac{1.25}{200} = 0.00625$$

$$1700 \text{ mg} \rightarrow 0.00625 \times 1700 = \textbf{£10.63}$$

## Questions

**Questions 1–14 are similar to those covered in worked examples 15.1–15.6.**

1   The enthalpy of combustion of benzene, $C_6H_6$, is $-3628 \text{ kJ mol}^{-1}$. Calculate the energy released when 18 g of benzene is burned.

2   The enthalpy of combustion of propane, $C_3H_8$, is $-2219 \text{ kJ mol}^{-1}$. Calculate the energy released when 36 g of propane is burned.

3   In an experiment, 18 g of hydrogen was burned and released 3750 kJ of energy. Calculate the experimental enthalpy of combustion of hydrogen (1 mol of hydrogen = 2 g).

4   In an experiment, 5 g of ethanol was burned and released 102 kJ of energy. Calculate the enthalpy of combustion of ethanol (1 mol of ethanol = 46 g).

5   The enthalpy of combustion of benzene (1 mol = 78 g) is $-3628 \text{ kJ mol}^{-1}$. Calculate the mass of benzene that should be burned to release 500 kJ of energy.

6   The enthalpy of combustion of ethyne (1 mol = 26 g) is $-1301 \text{ kJ mol}^{-1}$. Calculate the mass of ethyne that should be burned to release 145 kJ of energy.

7   A 10 g bottle of zinc(II) chloride cost £87. Calculate the cost for 0.34 g of zinc(II) chloride.

8   A 25 g bottle of magnesium cost £24. Calculate the cost for 123 g of magnesium.

9   8 g of breakfast cereal provided 0.09 mg of a vitamin. Calculate the mass of vitamin supplied by a 60 g serving of the cereal.

10   5 ml of a medicine contained 200 mg of the active ingredient. Calculate the mass of active ingredient supplied in a 300 ml bottle of the medicine.

11   360 g of $H_2O$ contains $1.204 \times 10^{25}$ molecules. Calculate the number of $H_2O$ molecules present in 0.85 g of $H_2O$.

12   A 1000 g sample of $CH_4$ contained $3.7625 \times 10^{25}$ molecules. Calculate the number of $CH_4$ molecules present in 1 mol of $CH_4$.

13   10 g of neon contained $6.02 \times 10^{23}$ neon atoms. Calculate the mass of 50 neon atoms.

14   1 kg of $NH_3$ molecules contained $3.54 \times 10^{25}$ molecules. Calculate the number of molecules present in 2 mol of $NH_3$.

**Questions 15–20 are similar to those covered in worked examples 15.7 and 15.8.**

15   The RDA of vitamin E is 16.5 mg. A single carrot can supply 0.66 mg of vitamin E. Calculate the number of carrots that could supply the RDA.

16   The RDA of vitamin C is 60 mg. A 50 ml serving of orange juice contained 48 mg of vitamin C. Calculate the volume of orange juice that could supply the RDA of vitamin C.

17  In an experiment, 8 g of ethyl propanoate was obtained from the reaction of 14 cm³ of propanoic acid with excess ethanol. The propanoic acid was obtained at a cost of £2.32 per 50 cm³. Calculate the cost of propanoic acid lrequired to make 1 kg of ethyl propanoate.

18  5 ml of a pain-relieving medication contained 100 mg of ibuprofen. Calculate the volume of medication that could be given to a 30 kg child if the maximum safe daily dose is 30 mg per kg.

19  The maximum level of oxalic acid permitted in the diet is estimated to be 0.14 mg per kg per day. A 10 ml serving of honey was found to contain 0.35 mg of oxalic acid. Calculate the maximum volume of honey that could be consumed safely in 1 day by a 55 kg woman.

20  10 ml of a cough remedy was found to contain 28 mg of codeine. The maximum safe dose of codeine is 360 mg in 24 hours. Calculate the maximum volume of cough remedy that could be consumed over 24 hours.

# Answers

## Chapter 1:

1  a)  $0.043\,s^{-1}$
   b)  $0.0026\,s^{-1}$
   c)  $0.0017\,s^{-1}$
   d)  $0.0083\,s^{-1}$
   e)  $0.0038\,s^{-1}$

2  a)  $83.33\,s$
   b)  $20.41\,s$
   c)  $909.09\,s$
   d)  $1769.91\,s$
   e)  $1481.48\,s$

3  a)  $0.37\,mol\,l^{-1}$
   b)  $50\,s$

4  a)  $25\,s$
   b)  $33\,°C$
   c)  $0.004\,s^{-1}$, $0.008\,s^{-1}$, $0.016\,s^{-1}$ and $0.032\,s^{-1}$. The rate doubles with each $10\,°C$ temperature rise.

5  a)  $0.32\,mol\,l^{-1}$
   b)  $40\,s$

6  a)  $25\,°C$
   b)  $25\,s$

7  a)  $62.5\,s$
   b)  $0.0084\,mol\,l^{-1}$

## Chapter 2:

1  a)  $110\,kJ$
   b)  $80\,kJ$

2  a)  $100\,kJ$
   b)  $70\,kJ$

3  a)  $70\,kJ$
   b)  $50\,kJ$

4  a)  (i)  $50\,kJ$
       (ii) $-40\,kJ$
   b)  $30\,kJ$

5  a)  b
   b)  a
   c)  A: a + b

6  a)  w
   b)  y
   c)  z

7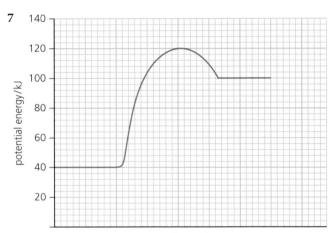

8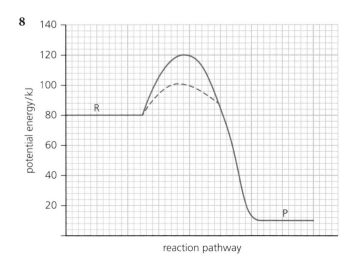

# Chapter 3:

1  32.8 g

2  0.02 mol

3  1.2 mol

4  8.55 g

5  0.0025 mol

6  59.6 g

7  0.3 mol

8  2.24 g

9  0.02 mol

10  48 g

11  26.9 g

12  1.5 mol

13  3.99 g

14  0.04 mol

15  147 g

16  0.8 mol

17  124.8 g

18  0.05 mol

19  31.8 g

20  0.03 mol

21  $0.2 \, mol \, l^{-1}$

22  0.3 mol

23  $0.2 \, mol \, l^{-1}$

24  $40 \, cm^3$

25  0.15 mol

26  $0.2 \, mol \, l^{-1}$

27  $250 \, cm^3$

28  32 g

29  $0.25 \, mol \, l^{-1}$

30  1.321 g

31  1.5 l

32  $40 \, cm^3$

33  0.5 l

34  7.98 g

35  $0.05 \, mol \, l^{-1}$

36  $0.05 \, mol \, l^{-1}$

37  $0.1 \, mol \, l^{-1}$

38  0.96 g

39  $0.2 \, mol \, l^{-1}$

40  $25 \, cm^3$

# Chapter 4:

1  0.5 g

2  12.7 g

3  0.6 g

4  3.5 g

5  11.5 g

6  0.4 g

7  7.29 g

8  2.2 g

9  66 g

10  1.6 g

11  1282 kg

12  37.5 kg

13  44.64 tonnes

14  9612 kg

15  672 kg

16  $2.76 \times 10^3 \, kg$

17  $3.52 \times 10^4 \, kg$

18  3.68 tonnes

19  $1.9 \times 10^5 \, kg$

20  $8.8 \times 10^3 \, kg$

# Chapter 5:

1  a)  Oxygen

   b)  56 g

2  a)  Methane

   b)  12.38 g

3  a)  Lead

   b)  0.05 g

4  a)  Calcium carbonate

   b)  4.4 g

5  a)  Potassium iodide

   b)  11.06 g

6  15.23 g

7  20.25 g

8  1.1 g

9  0.4 g

10 a)  0.02 mol of aluminium nitrate and 0.08 mol of sodium hydroxide

      1 mol of aluminium nitrate will react with 3 mol of sodium hydroxide.

      0.02 mol of aluminium nitrate will react with 0.06 mol of sodium hydroxide.

      Therefore, sodium hydroxide is in excess.

   b)  1.56 g

   c)  0.02 mol

11 a)  0.0018 mol of silver(I) nitrate and 0.0012 mol of magnesium chloride

      0.0018 mol of silver(I) nitrate reacts with 0.0009 mol of magnesium chloride.

      Therefore, magnesium chloride is in excess.

   b)  0.258 g

   c)  $3 \times 10^{-4}$ mol

12 a)  0.004 mol of sulfuric acid and 0.005 mol of barium chloride

      0.004 mol of sulfuric acid reacts with 0.004 mol of barium chloride.

      Therefore, barium chloride is in excess.

   b)  0.934 g

13 a)  0.05 mol of ammonium sulfate and 0.06 mol of sodium hydroxide

      0.05 mol of ammonium sulfate reacts with 0.10 mol of sodium hydroxide.

      Therefore, ammonium sulfate is in excess.

   b)  1.02 g

14 a)  0.05 mol of iron(II) sulfide and 0.08 mol of hydrochloric acid

      0.05 mol of iron(II) sulfide reacts with 0.10 mol of hydrochloric acid.

      Therefore, iron sulfide is in excess.

   b)  1.36 g

15 a)  0.02 mol of copper(II) oxide and 0.012 mol of sulfuric acid

      0.02 mol of copper(II) oxide will react with 0.02 mol of sulfuric acid.

      Therefore, copper(II) oxide is in excess.

   b)  0.636 g

   c)  1.92 g

# Chapter 6:

1  268.8 litres

2  11.2 litres

3  4.48 litres

4  2.24 litres

5  0.22 mol

6  0.036 mol

7  0.0045 mol

8  0.022 mol

9  25 litres mol$^{-1}$

10 22.5 litres mol$^{-1}$

11 22.4 litres mol$^{-1}$

12 0.0236 g

13 3.24 litres

14 100 litres

15 184.78 g

16 0.045 litres

17  24.3 litres

18  1101 g

19  0.025 litres

20  24.31 litres mol$^{-1}$

21  23.045 litres mol$^{-1}$

22  22.45 litres mol$^{-1}$

23  23.08 litres mol$^{-1}$

24  24 litres mol$^{-1}$

25  71 g; $Cl_2$

26  30 g; $C_2H_6$

27  44 g; $CO_2$

28  a)  24 litres mol$^{-1}$

　　b)  38

29  15.82

30  31.8

## Chapter 7:

1  a)  $O_2$

　　b)  100 cm$^3$ $O_2$ and 100 cm$^3$ $CO_2$

2  a)  CO

　　b)  40 cm$^3$ $CO_2$ and 10 cm$^3$ CO

3  60 l $CO_2$ and 40 l $O_2$

4  1200 l $CO_2$ and 3200 l $O_2$

5  400 cm$^3$ $CO_2$, 600 cm$^3$ $H_2O$ and 1300 cm$^3$ $O_2$

6  100 l $N_2$, 200 l $H_2O$ and 300 l $O_2$

7  600 l $CO_2$ and 1025 l $O_2$

8  a)  $2.4 \times 10^5$ l $CO_2$

　　b)  $5.2 \times 10^5$ l $H_2$

9  $5 \times 10^5$ l $CCl_4$, $5 \times 10^5$ l $S_2Cl_2$ and $10^6$ l $Cl_2$

10  $4.5 \times 10^5$ l

## Chapter 8:

1  9.9 litres

2  1.1 litres

3  320 litres

4  352 litres

5  11.89 g

6  2.2 litres

7  20 litres

8  3636.36 g

9  3.52 litres

10  234 kg

## Chapter 9:

1  75%

2  40%

3  75%

4  75%

5  80%

6  60%

7  70%

8  60%

9  11.17 g

10  13.44 g

11  0.45 g

12  33.48 g

13  5.75 g

14  $7.14 \times 10^4$ kg

15  27.6 kg

16  $7.20 \times 10^4$ kg

17  6512 kg

18  666 kg

**19 a)** 4.185 kg

**b)** 90%

**20 a)** $3.99 \times 10^4$ kg

**b)** 60%

# Chapter 10:

**1** 45.8%

**2** 56%

**3** 17.65%

**4** 77.7%

**5** 20%

**6** 100%

**7** 51%

**8** 100%

**9** 56%

**10** 80%

**11** $CO + 2H_2 \rightarrow CH_3OH$

Total mass of reactants = CO (12 + 16) + 2H$_2$ (2 × 2) = 32

Total mass of product = CH$_3$OH = 32

atom economy = $\dfrac{\text{mass of desired product(s)}}{\text{total mass of reactants}} \times 100$

atom economy = $\dfrac{32}{32} \times 100 = 100\%$

**12** 69%

# Chapter 11:

**1** 125.4 kJ

**2** 8.36 kJ

**3** 4.18 kJ

**4** 1.25 kJ

**5** 95.69 °C

**6** 69.85 °C

**7** 36.04 °C

**8** 47.06 °C

**9** 0.239 kg

**10** 0.219 kg

**11** 3.34 kJ

**12** 2.09 kJ

**13** 39.71 kJ

**14** $-188.1\,\text{kJ mol}^{-1}$

**15** $-334.4\,\text{kJ mol}^{-1}$

**16** $-167.2\,\text{kJ mol}^{-1}$

**17** $-355.3\,\text{kJ mol}^{-1}$

**18** $-668.8\,\text{kJ mol}^{-1}$

**19** $-836\,\text{kJ mol}^{-1}$

**20** $-1337.6\,\text{kJ mol}^{-1}$

**21** $-2090\,\text{kJ mol}^{-1}$

**22** $-2664.75\,\text{kJ mol}^{-1}$

**23** 0.42 g

**24** 0.34 g

**25** 10.66 °C

**26** 33.15 g

**27** 13.48 g

**28** 3.11 litres

**29** 2.97 litres

**30** 21.71 °C

# Chapter 12:

**1** $-545\,\text{kJ mol}^{-1}$

**2** $-102\,\text{kJ mol}^{-1}$

**3** $-9\,\text{kJ mol}^{-1}$

**4** $-694\,\text{kJ mol}^{-1}$

**5** $-1680\,\text{kJ mol}^{-1}$

**6** $-608\,\text{kJ mol}^{-1}$

**7** $-124\,\text{kJ mol}^{-1}$

**8** $-45\,\text{kJ mol}^{-1}$

**9** $-169\,\text{kJ mol}^{-1}$

**10** $-54\,\text{kJ mol}^{-1}$

**11** $226.5\,\text{kJ mol}^{-1}$

**12** $216.3\,\text{kJ mol}^{-1}$

**13** $384.16\,\text{kJ mol}^{-1}$

**14** $163\,\text{kJ mol}^{-1}$

**15** $840\,\text{kJ mol}^{-1}$

# Chapter 13:

1   $-891\,kJ\,mol^{-1}$

2   $-1561\,kJ\,mol^{-1}$

3   $-86\,kJ\,mol^{-1}$

4   $-2222\,kJ\,mol^{-1}$

5   $-129\,kJ\,mol^{-1}$

6   $-1368\,kJ\,mol^{-1}$

7   $-484\,kJ\,mol^{-1}$

8   $-306\,kJ\,mol^{-1}$

9   $-3271\,kJ\,mol^{-1}$

10   $226\,kJ\,mol^{-1}$

11   $-425\,kJ\,mol^{-1}$

12   $-493\,kJ\,mol^{-1}$

13   $-136\,kJ\,mol^{-1}$

14   $-79\,kJ\,mol^{-1}$

15   $-312\,kJ\,mol^{-1}$

16   $-412\,kJ\,mol^{-1}$

17   $377\,kJ\,mol^{-1}$

18   $776\,kJ\,mol^{-1}$

19   $-409\,kJ\,mol^{-1}$

20   $-5120\,kJ\,mol^{-1}$

# Chapter 14:

1   $0.5\,mol\,l^{-1}$

2   $16\,cm^3$

3   $6.91\,g$

4   $0.3\,litres$

5   $0.8\,mol\,l^{-1}$

6   $4.15\,g$

7   $0.182\,mol\,l^{-1}$

8   $1.17\,g$

9   $0.094\,mol\,l^{-1}$

10   $0.2\,mol\,l^{-1}$

11   $2\,mol\,l^{-1}$

12   $100\,cm^3$

13   $0.32\,mol\,l^{-1}$

14   $2.5\,mol\,l^{-1}$

15   a)   $0.02\,mol$

   b)   $0.04\,mol\,l^{-1}$

16   $50\,cm^3$

17   $16\,cm^3$

18   $0.0625\,mol\,l^{-1}$

19   $40\,cm^3$

20   $0.08\,mol\,l^{-1}$

21   $0.05\,mol\,l^{-1}$

22   $0.4\,mol\,l^{-1}$

23   $18\,cm^3$

24   $20\,cm^3$

25   $38.4\,cm^3$

# Chapter 15:

1   $-837.23\,kJ\,mol^{-1}$

2   $-1815.54\,kJ\,mol^{-1}$

3   $-416.67\,kJ\,mol^{-1}$

4   $-938.4\,kJ\,mol^{-1}$

5   $10.75\,g$

6   $2.90\,g$

7   £2.96

8   £118

9   $0.675\,mg$

10   $12\,000\,mg$

11   $2.84 \times 10^{22}$

12   $6.02 \times 10^{23}$

13   $8.31 \times 10^{-22}\,g$

**14** $1.204 \times 10^{24}$

**15** 25

**16** 62.5 ml

**17** £81.20

**18** 45 ml

**19** 220 ml

**20** 128.57 ml